MW01517901

DARWIN AND THE MEMORY OF THE HUMAN

When the young Charles Darwin landed on the shores of Tierra del Fuego in 1832, he was overwhelmed: nothing had prepared him for the sight of what he called "an untamed savage." The shock he felt, repeatedly recalled in later years, definitively shaped his theory of evolution. In this original and wide-ranging study, Cannon Schmitt shows how Darwin and other Victorian naturalists transformed such encounters with South America and its indigenous peoples into influential accounts of biological and historical change. Redefining what it means to be human, they argue that the modern self must be understood in relation to a variety of pasts – personal, historical, and ancestral – conceived of as savage. Schmitt reshapes our understanding of Victorian imperialism, revisits the implications of Darwinian theory, and demonstrates the pertinence of nineteenth-century biological thought to current theorizations of memory.

CANNON SCHMITT is Associate Professor of English at the University of Toronto.

Nineteenth-century British literature and culture have been rich fields for interdisciplinary studies. Since the turn of the twentieth century, scholars and critics have tracked the intersections and tensions between Victorian literature and the visual arts, politics, social organization, economic life, technical innovations, scientific thought – in short, culture in its broadest sense. In recent years, theoretical challenges and historiographical shifts have unsettled the assumptions of previous scholarly synthesis and called into question the terms of older debates. Whereas the tendency in much past literary critical interpretation was to use the metaphor of culture as "background," feminist, Foucauldian, and other analyses have employed more dynamic models that raise questions of power and of circulation. Such developments have reanimated the field. This series aims to accommodate and promote the most interesting work being undertaken on the frontiers of the field of nineteenth-century literary studies: work which intersects fruitfully with other fields of study such as history, or literary theory, or the history of science. Comparative as well as interdisciplinary approaches are welcomed.

A complete list of titles published will be found at the end of the book.

THE
LONDON SKETCH BOOK.

PROF. DARWIN.

This is the ape of form.
Love's Labor Lost, act 5, scene 2.

Some four or five descents since.
All's Well that Ends Well, act 3, sc. 7.

"Prof. Darwin," *Figaro's London Sketch Book of Celebrities*,
February 18, 1874. Color lithograph by Faustin Betbeder. Courtesy
the Wellcome Library, London.

$ 105.95

DARWIN AND THE MEMORY OF THE HUMAN

Evolution, Savages, and South America

CANNON SCHMITT

University of Toronto

CAMBRIDGE
UNIVERSITY PRESS

CAMBRIDGE UNIVERSITY PRESS
Cambridge, New York, Melbourne, Madrid, Cape Town, Singapore, São Paulo, Delhi

Cambridge University Press
The Edinburgh Building, Cambridge CB2 8RU, UK

Published in the United States of America by Cambridge University Press, New York

www.cambridge.org
Information on this title: www.cambridge.org/9780521765602

First published 2009

Printed in the United Kingdom at the University Press, Cambridge

A catalogue record for this publication is available from the British Library

Library of Congress Cataloguing in Publication data
Schmitt, Cannon.
Darwin and the memory of the human : evolution, savages, and
south america / Cannon Schmitt.
p.   cm. – (Cambridge studies in nineteenth-century literature and culture)
Includes bibliographical references and index.
ISBN 978-0-521-76560-2 (hardback)   1. Nature in literature.   2. Human evolution in
literature.   3. Human evolution – Philosophy.   4. Darwin, Charles, 1758–1778 –
Travel – South America.   5. Wallace, Alfred Russel, 1823–1913 – Travel – South
America.   6. Kingsley, Charles, 1819–1875 – Travel – South America.   7. Hudson, W. H.
(William Henry), 1841–1922 – Travel – South America.   8. English literature – 19th century –
History and criticism.   I. Title.   II. Series.
PR468.N3S36 2009
820.9′36–dc22
2009010964

ISBN 978-0-521-76560-2 hardback

In Memoriam
Tabitha Beckett Chesnut, 1968–1998
James Doss Chesnut, 1941–1975
Bryant Lester Yeomans, 1921–1999

Contents

Illustrations

Acknowledgements

For insights, provocations, disagreements, suggestions, references, corrections, bemusement, excitement, and more – which is just to say for all the sundry manifestations of that most important of collegial virtues, engagement – I thank David Aers, Robert Aguirre, Tim Alborn, Emily Allen, Suzy Anger, Srinivas Aravamudan, Anjali Arondekar, Ian Baucom, Sarika Chandra, Ian Duncan, Mary Favret, Dino Felluga, Jonathan Flatley, Ross Forman, Elaine Freedgood, Jared Gardner, Richard Grusin, Elaine Hadley, Nancy Henry, Beth Hewitt, Neville Hoad, Anne Humpherys, Gerhard Joseph, Amy King, John Kucich, Donna Landry, Gerald MacLean, Dwight McBride, Andrew Miller, Michael Moses, John Plotz, Maureen Quilligan, Ben Schmidt, Alan Schrift, Peter Stallybrass, Marianna Torgovnick, and Priscilla Wald. Lee Sterrenburg introduced me to Darwin in the first place. New colleagues at the University of Toronto have extended an unusually warm welcome to their (now, happily, our) vibrant intellectual community; I offer them all my gratitude. For interest in and help with this project in particular, Alan Bewell, Michael Cobb, Brian Corman, Deidre Lynch, Jill Matus, Paul Stevens, and Sarah Wilson merit special mention. Terrific students at Duke University, Wayne State University, and the University of Toronto contributed more than they know, especially Kristine Danielson, Ryan Dillaha, Nihad Farooq, Melissa Free, Jacques Khalip, Justin Prystash, Sarah Ruddy, and Michael Schmidt. I refined several claims in response to sharp questions from audiences at the University of Pennsylvania, the City University of New York Graduate Center, the State University of New York at Binghamton, the University of Chicago, and a number of conferences. At Cambridge University Press, Gillian Beer and Linda Bree showed a heartening enthusiasm for the project, and Maartje Scheltens set a high standard of efficiency and professionalism. Anonymous readers for the Press made the book better than it would have been without them.

My institutional debts are legion. A Harris Fellowship from Grinnell College and a Career Development Chair from Wayne State University provided invaluable time; grants from the National Endowment for the Humanities, Grinnell College, Duke University, the Trent Memorial Foundation, Wayne State University, and the University of Toronto, crucial funding; and the Newberry Library, Chicago, a congenial place to work at an early stage. The Bibliothèque nationale de France, the British Library, the New York Public Library, the Wellcome Library, London, and various other such astounding storehouses contributed their share, but I don't know how I could have finished the book without the history of science collections at the Thomas Fisher Rare Book Library, University of Toronto – above all the Darwin Collection, initially assembled by Richard B. Freeman and assiduously developed over the years by the Fisher's director, Richard Landon.

An earlier version of chapter 1 appeared as "Darwin's Savage Mnemonics," *Representations* 88 (Fall 2004): 55–80. © 2005 by the Regents of the University of California. Permission to reprint is gratefully acknowledged.

One of my beloved and influential aunts, Saralyn Chesnut, tells a story about how in childhood her older brothers, one of whom was my father, liked to tease her with vocabulary she hadn't yet learned. "You have ancestors," they would say accusingly – to which she would reply, scandalized, "No I don't!" At the end of writing a book about, among other things, the scandal (but also the pleasure) of having to acknowledge certain ancestors and relatives, I am put in mind of how much is due members of my family, both those no longer with us and those who remain. It is to the memory of three of them, my brother, Tabby Beckett Chesnut, my father, James Doss Chesnut, and my maternal grandfather, B. L. Yeomans, that the book is dedicated. Helen Chesnut, my paternal grandmother, leaves a void as well. Much love, thanks, and appreciation go to Saralyn; Rebecca Yeomans, my other beloved and influential aunt; Sally Yeomans, my maternal grandmother; and my indefatigably loving and supportive stepfather, Jack Schmitt, and mother, Zebe Schmitt. Dana Seitler emboldened me to write the book I wanted to rather than the one I thought I should. Beyond that, and beyond being my toughest critic and most reliable friend, she continues to reveal the ways in which this seemingly all-too-explored world brims with possibility. Finally, for the wonder of our son Beckett there are no words. As the figures I write about find themselves doing so often, I will simply have to take refuge in observing that some things, the most momentous, are ineffable – and leave it at that.

Introduction

"No tropical country is more interesting than South America," declares Grant Allen in his introduction to James Rodway's *In the Guiana Forest: Studies of Nature in Relation to the Struggle for Life* (1894).[1] This judgement has nothing to do with the continent's fabled gold or the ruins of its pre-Columbian civilizations. For Allen, who finds himself in close accord with Rodway's subtitle, South America stands as the most interesting of tropical countries because of its natural productions, particularly insofar as those productions shed light on "the struggle for life" – that is, on evolutionary history and theory. Asia, Africa, and Australia also compel attention in that each, according to Allen, exemplifies an extremely early period in the development of life. But South America possesses unique appeal for professional biologists as well as amateur students of the natural world because "it preserves for us, as in amber, numberless intermediate stages" of evolutionary change.[2] Moreover, and despite the metaphor of fossilization encoded in the reference to amber, the South American representatives of those intermediate stages so fascinate because they take the form of actually existing flora and fauna rather than petrified insects or dry shards of bone. From ferns and flowering plants to birds and mammals, including those mammals Allen and Rodway refer to as "savages," South America's extant biota comprises a living archive in which the evolutionary past of the middle distance may be encountered in all its teeming glory.[3]

I begin with Allen's paean to the New World tropics because it constitutes an arresting example of a more widespread phenomenon: the Victorian depiction of South America as an immense site of memory or *lieu de mémoire*.[4] The neologism belongs to Pierre Nora, who coins it in a meditation on the disappearance of organic, self-sustaining "environments of memory" and their replacement by artificially constructed and maintained "sites." "*Lieux de mémoire* originate," he writes, "with the sense that there is no spontaneous memory, that we must deliberately create archives, maintain anniversaries, organize celebrations, pronounce eulogies, and notarize bills because such activities no longer occur naturally."[5] Nora argues that, lacking those rituals that once effortlessly joined one day, week, and year to the next, and in so doing ensured the existence of memory as a vibrant, felt continuity, we must now fabricate sites for memory to reside, sites that function "to stop time, to block the work of forgetting … to materialize the immaterial."[6] Although Nora dates the advent of *lieux de mémoire* to the twentieth century, and although a continent cannot be spoken of as fabricated in quite the same sense that a museum or library may be, South America provided for the Victorians a place for just such stopping, blocking, and materializing, a place that memorialized the past in such a way as to make it available in the present.

Most immediately and obviously, South America memorialized empire, its failures as well as its unrealized possibilities. Nineteenth-century Britons who witnessed the remnants of Spanish and Portuguese empires in the Americas often fell into modes of imperial lamentation, deploring what they perceived as ruinous Iberian policy while plangently imagining what might have been had only the British arrived earlier, remained longer, fought harder. Consider the young Thomas Henry Huxley, who, during a brief stay in Rio de Janeiro en route to the Pacific, indulges in a flight of historical fancy remarkable only because so commonplace among his compatriots: "A few of the hungry Saxon millions now famishing in England, had they possession of such a country as this, and the Brazilians extirpated, might found a second Indian Empire."[7] But the particulars of Allen's contentions, which locate the value of South America in its plants and animals, reveal another and stranger valence of the continent as a repository of the past, one more in line with Huxley's eventual vocation as an indefatigable promoter of Darwinism than his callow geopolitical musings. Like geological deposits whose deepening layers of sediment correspond to ever-receding periods of the earth's development (a simile recently available in the nineteenth century), South America was for the Victorians a space encoding the passage of evolutionary time – a rich assemblage of living beings that, despite continuing to thrive in the present, nonetheless recalled the history of life on the planet.[8]

In *Darwin and the Memory of the Human: Evolution, Savages, and South America*, I investigate the Victorian engagement with South America as a *lieu de mémoire*. Further, because of the paramount role of natural history and evolutionary theory in constituting that continent as such a site, and because of the corresponding place of South America in the genealogy of the theory of evolution, I do so by attending to the works of a group of remarkable and widely influential naturalists who traveled there and wrote about what they saw: Charles Darwin; Alfred Russel Wallace, independent co-founder of the theory of evolution by means of natural selection; Charles Kingsley, novelist, opponent of the Oxford Movement, and popularizer of science; and W. H. Hudson, memoirist, novelist, and ornithologist. All four men in their different ways encounter South America as and through memory. All parlay that encounter into narratives about savagery and civility, race and the origins of humanity – narratives they find at times consoling, at times disturbing. Many of those narratives take shape around indigenous South Americans or "savages" who, according to the dictates of the comparative method, are understood at once as living memories (ancestors who have somehow persisted into the present) and instigations to remembering (reminders of what humans once were).[9] Others concern themselves more centrally with the recollected exploits of European precursors, embracing belatedness as a privileged form of encounter and an enabling narrative trope. Still others involve a return to and reenactment of memories individual as well as national and "racial." All elaborate what I call a savage mnemonics: a form of memory that redefines what it is to be human (as well as modern, civilized, and British) in relation to the past, and specifically those pasts – historical, cultural, personal, and, above all, evolutionary – conceived of as savage.

At its most ambitious, *Darwin and the Memory of the Human* seeks to tell a story about the invention of a new and enduring human subject in the nineteenth century – not the only such story, to be sure, but a crucial and heretofore neglected one nonetheless. The human in question was decisively modern because thoroughly biological and "natural": an animal among animals, a product of inexorable laws working themselves out over the course of profane, sublunary time no less than beetles or monkeys. As Darwin puts it in the final sentence of *The Descent of Man, and Selection in Relation to Sex* (1871), fusing his most daring assertion with some of the key evidence for its validity: "Man still bears in his bodily frame the indelible stamp of his lowly origin."[10] But this natural entity was, nevertheless, also an invention. This is the case not simply because, like all notions of the human, the modern human-as-animal was and remains a cultural artifact, but also in the sense that, more precisely and interestingly, it came into being by means of a particular technology: memory. This human emerged, in Darwin's work and elsewhere, by way of how it was remembered. Accordingly, one burden of my argument has to do with anatomizing the ways that memory, as a technology of the human, enabled the invention of the human as natural.

By using the term "technology," I hope to signal the historicity of what goes by the name of memory as well as to prepare the ground for a recovery of some of its varied forms and purposes at different historical moments. John Frow writes, in a somewhat different context, that "[t]o speak of memory as *tekhnè*, to deny that it has an unmediated relation to experience, is to say that the logic of textuality by which memory is structured has technological and institutional conditions of existence."[11] It is therefore also to say that what counts as memory changes as those conditions of existence change. The writers I study elaborated and deployed a cluster of closely related forms of memory in the service of postulating "lowly origin[s]" that could be recollected: corporeal or unconscious memory, which imagines that the past inheres in the body of the individual; "racial" memory, which posits a faculty of recall that coheres around and lends coherence to an ancestry conceived in terms of race and nation; and something that might be named nature-memory, which understands the natural world in its entirety as a vast biological memorial. For "civilized" men such as these writers saw themselves to be, establishing their humanity entailed establishing continuity with the distant past achieved by remembering their own animal, inhuman, and savage origins. In the process, the state of being civilized itself came to be defined as the capacity to remember savage beginnings at once surpassed and (in the individual and elsewhere) preserved. In other words, ones that Bruno Latour might use: the human is a purity that comes into being by the production and disavowal of hybrids (the hybrid animal/human as well as the hybrid ancient/modern).[12] Thus did an epistemology indebted to a certain mnemotechnics – to know the human was to remember, and to remember specifically an outmoded or incipient version of oneself – imply an ontology riven by paradox – the human is the animal that remembers it no longer is an animal, the savage whose sole remaining connection to savagery is in memory.

In this book, then, I document some of the conditions of the formation of a modern version of the human, one that remains very much with us: the human as the subject of evolution.[13] If, despite the efforts of certain US boards of education and the proponents of the current incarnation of creationism, so-called "intelligent design," the truth of this version of ourselves now seems merely self-evident, its implications mundane rather than startling, this may be because of the degree to which it has come down to us as an abstraction.[14] At its inception, however, this conception of the human emerged from the confluence of various kinds of specificity with consequences for our view of the Victorians, our approach to them and their textual productions, and our sense of ourselves.

In the works of the writers and thinkers I study, one such specificity is geographical: South America provides them with the locus for their reinvention of the human. The centrality of that continent in their accounts necessitates, among other things, a rethinking of received notions about Victorian racial ideology, so often understood to coalesce around a black/white or east/west binary. Falling outside of and to some degree confounding both binaries, South American "savages" evidence the existence of a logic of alterity that is at the same time a logic of similitude, accommodating complex negotiations of difference and likeness across not only spatial but also temporal distance. Similarly, while Victorians traveling in and writing about South America often have recourse to colonial or imperial modalities of thought, analyses resulting from the straightforward invocation of the rubric "empire," while necessary and illuminating, are insufficient in themselves, even in their metaphorical or analogical application. New models of transnationalism are required, models that incorporate even as they revise familiar versions of Victorian Britain's relations with the rest of the globe.

Another consequential specificity of the reinvention of the human with which I am concerned has to do with a scientific practice and its associated conventions: natural history, which, over the course of the nineteenth century, gave rise to and was in turn transformed by evolutionary theory.[15] The particularities of natural history as a conceptual, discursive, and practical regime definitively inflected the notion of the human that took shape in connection with nineteenth-century South America. Most importantly, the catholicity of the natural-historical enterprise at the beginning of the Victorian period ensured that not only plants and animals but also minerals and cultures fell within the purview of the scientific traveler.[16] The natural-historical collection could and did contain stuffed bird skins, pinned insects, pictographs broken from rock walls, clothing, cooking implements, and human remains.[17] Out of this congeries came museum displays, travel narratives, scientific monographs, and, eventually, claims about human kinship with what must now be called other animals.

Finally, neither natural history nor Victorian South America could have taken the shape they did without the aid of various kinds of memory. While our more familiar definitions of "memory" still recognize some of those – most commonly, for instance, individual recollection with its attendant feelings of belatedness or nostalgia – others are lost to us. Nora notes that among the "costs of the historical metamorphosis of memory" may be counted "a wholesale preoccupation with the individual psychology of remembering."[18] As a result, anything falling outside that psychology risks losing its resemblance to memory altogether. But the body as a retainer of the archaic past, the growth of the nation as collective commemoration, the natural world as all-encompassing archive: for the Victorians these all counted as memory, and a key contention of this book has to do not simply with their existence as such but with their indispensability to the development of what have proven to be more enduring conceptual formations, evolutionary theory and the human as the subject of evolution chief among them.

SOUTH AMERICOMANIA

In 1824, British Foreign Secretary George Canning asserted: "Spanish America is free and if we do not mismanage our affairs sadly, she is English."[19] Written near the end of a series of independence movements that delivered Mexico and most of Central and South America from Spanish and Portuguese rule, Canning's assertion rings with enthusiasm for the opening up of territories that had been largely closed since the sixteenth century. British troops fought beside Simón Bolívar and José de San Martín in the struggle against Spain, and for many Britons the newly independent nations of the Americas represented a victory of national self-determination over despotism. At the same time, those nations also appealed to the British for the same reasons that their own considerable and growing imperial holdings in North America, the Caribbean, Asia, and Africa did: they promised heretofore untapped markets, mineral and agricultural wealth, and a wide field for the aspirations of British travelers of all sorts – not only the naturalists with whom I am primarily concerned but also mercenaries, businessmen, bankers, immigrants, mining engineers, and mountaineers. The flood of investment, trade, and travel that poured in from the late 1820s on made Britain the dominant foreign power in Latin America until it ceded that role to the United States in the decade following the First World War. In Leslie Bethell's pithy formulation, "The nineteenth century was the 'British century' in Latin America."[20]

Bethell's statement is not reversible; the nineteenth century can hardly be described as the "Latin-American century" in Britain. Still, the Victorians did maintain a continuous and often acute level of interest in the region attested to by, among other things, numerous and diverse texts: personal histories of the wars of independence, such as Thomas Cochrane's *Narrative of Services in the Liberation of Chile, Peru and Brazil* (1860);

guides to immigration, including Henry C. R. Johnson's *A Long Vacation in the Argentine Alps, or Where to Settle in the River Plate States* (1868); anti-slavery tracts such as Edward Wilberforce's *Brazil Viewed through a Naval Glass; with Notes on Slavery and the Slave Trade* (1856); travel narratives like Lady Florence Caroline Dixie's *Across Patagonia* (1880) and Edward Whymper's *Travels Amongst the Great Andes of the Equator* (1891); accounts of mining ventures, including Sir Richard Francis Burton's *Explorations of the Highlands of Brazil; with a Full Account of the Gold and Diamond Mines* (1869) and Alexander James Duffield's *Peru in the Guano Age: Being a Short Account of a Recent Visit to the Guano Deposits with Some Reflections on the Money They Have Produced and the Uses to Which It Has Been Applied* (1877); and novels, from R. M. Ballantyne's *Martin Rattler; or, A Boy's Adventure in the Forests of Brazil* (1858) to Joseph Conrad's *Nostromo* (1904).[21]

Although very little of Latin America was actually in British possession during this period – on the mainland of South America, only the small settlement of British Guiana – much of the discourse on the region features the themes, figures of speech and thought, and narrative structures characteristic of writings on actual or prospective parts of the empire. In particular, like other non-European spaces, Latin America is often represented as the locus of the primitive or the atavistic. In *Imperial Leather: Race, Gender, and Sexuality in the Colonial Contest*, Anne McClintock names this trope "anachronistic space," a discursive strategy by means of which "the stubborn and threatening heterogeneity of the colonies was contained and disciplined not as socially or geographically different from Europe and thus equally valid, but as *temporally* different and thus as irrevocably superannuated by history."[22] For a stunningly literal instance of the depiction of South America as an anachronistic space, one need look no further than Sir Arthur Conan Doyle's 1912 *The Lost World*, the source of the title for this section of my introduction. One of Doyle's characters, Lord Roxton, is diagnosed as a "South Americomaniac": "He could not speak of that great country without ardour, and this ardour was infectious."[23] The novel follows Roxton and other members of a small group of adventurers as they push their way to the top of a rainforest plateau to discover not only ape-men who appear to provide the missing link between humans and other primates but also living dinosaurs, one specimen of which, a pterodactyl, they bring back with them as proof of what they have seen. (It escapes at novel's end, a prehistoric monstrosity roaming the skies above twentieth-century London.) Doyle's narrator describes Professor Challenger, the leader of this expedition through space and time, as "a Columbus of science who has discovered a lost world," articulating the British relation to Latin America as that between (admittedly belated) discoverer and heretofore unknown, or once known and since forgotten ("lost"), land.[24]

Although Doyle's dinosaurs obviously constitute a wildly exaggerated conception of what it meant for South America to be an anachronistic space, the difference from apparently more sober accounts is one of degree rather than kind. The British botanist and colonial administrator Everard im Thurn, for instance, whose writings were a key source for Doyle, observes of the area around Roraima, the Venezuelan *tepuy* or table-top mountain that served as the model for the lost world: "To the ethnologist also the district will prove interesting; for it is so remote and unexplored that the Arecuna Indians, who chiefly inhabit it, are in a very unusually primitive condition."[25] Doyle's novel, im Thurn's ethnography, and a host of other texts represent Latin America as an anachronistic space in McClintock's sense, out of step with Britain and so subject to "discovery" and conquest in the name of modernization. The nearly complete absence of British-controlled territory in the region, however, suggests that if there was a conquest in history to match that invoked in texts it rarely took the shape of formal occupation. Britain's only significant nineteenth-century attempt to increase its holdings in South America was the failed invasion of the Río de la Plata in 1806–07. Searching for a model with which to describe the relations between Victorian Britain and the new nations of Central and South America, economic and political historians have had recourse to terms such as "informal empire," first suggested by Ronald Robinson and John Gallagher, or "business imperialism," coined by D. C. M. Platt, both precursors of what would come to be known as neo-colonialism.[26] In *British Imperialism: Innovation and Expansion, 1688–1914*, P. J. Cain and A. G. Hopkins go so far as to consider South America "the crucial regional test" for their hypothesis that empire, at least in Britain's case, is most usefully and accurately understood as an economic rather than exclusively politico-military phenomenon.[27] For Cain and Hopkins, the British empire comprised not only areas under the control of the Crown but all those parts of the world in which the influence of British financial interests attenuated without necessarily abrogating local sovereignty.[28]

The difference between outright conquest and the British role in Latin America seems to have been grasped by Doyle, who is careful in *The Lost World* to describe Professor Challenger as a "Columbus of *science*." That specification, which borrows from but simultaneously reinflects the mythology of discovery, emphasizes the degree to which the exploration of the once more new continent of South America took place under the aegis of the expansion of knowledge rather than territory. As Nancy Stepan observes, "The New World political independence movements in the early nineteenth century open[ed] up the South American tropical regions to an intellectual and visual rediscovery."[29] In the vanguard of those undertaking the voyage from Britain to South America during this period, not only to the tropics but to the Andean highlands and Patagonian plains as well, were naturalists for whom that voyage meant an opportunity to study a landscape, flora, and fauna entirely different from what might be met with in Europe. With interests ranging from the botanical and geological to the entomological and ethnological, they represent themselves as driven by an epistemophilia in which the quest for natural-historical knowledge is inseparable from wonder and appreciation but unrelated to usurpation, domination, or expansionism.

Despite such self-representations, these naturalists were necessarily implicated in empire both practically and ideologically. In practical terms, most of the expeditions they mounted were made possible by a national-imperial infrastructure involving, for instance, transportation on British naval vessels and hospitality extended by resident British diplomats, merchants, and settlers. Moreover, they were also always on the alert for potentially valuable commodities – rubber, quinine, and coffee, to mention only the most obvious candidates.[30] Ideologically, too, these naturalists participated in the work of empire, in part by denying that they were doing so. In *Imperial Eyes: Travel Writing and Transculturation*, Mary Louise Pratt names the dynamic by which eighteenth- and nineteenth-century European travelers to the non-European world explicitly distanced themselves from an imperial enterprise in which they were nevertheless ineluctably engaged with the term "anti-conquest," a representational gambit "whereby European bourgeois subjects seek to secure their innocence" – that is, their absolute difference from the conquistadors, slavers, and settlers who have come before them – "in the same moment as they assert European hegemony." The naturalist in particular, argues Pratt, constituted "a utopian image of a European bourgeois subject simultaneously innocent and imperial, asserting a harmless hegemonic vision."[31] The quest for natural-historical knowledge, while in principle distinct from the imperial project, often enabled the exploration of new territory, advanced the possibility of exerting control over that territory, and extolled the desirability of doing so. In sum, as D. Graham Burnett has shown in his study of the writings of farmer, amateur ichthyologist, and ethnologist William Hillhouse and cartographer Robert Schomburgk, nineteenth-century British science and empire are often so inter-involved that "it is demonstrably difficult to separate 'scientific' exploration from colonial reconnaissance and colonial administration."[32]

Made in connection with explorers of the highlands of British Guiana, Burnett's observation also speaks to the work of British travelers in the rest of South America. But what modifications might be required in a context in which there was no properly colonial reconnaissance or administration to speak of? Or, to put the question more bluntly: what difference does informality make? One answer might be that Victorian scientific endeavor in South America was in many instances tied to goals more consistent with the expansion of capitalism and the securing of political hegemony than with seizure or administration of territory. Along these lines, Robert Aguirre, writing about Great Britain's engagement with Mexico and Central America in the nineteenth century, notes that scientific texts, panoramas, museum displays, and other "cultural forms engendered by this engagement … lent crucial ideological support to the practice of informal imperialism, shaping an audience receptive to the influx of British power in the region."[33] This argument, *mutatis mutandis*, closely resembles that of Edward Said, who asserts that "[i]n British culture … one may discover a consistency of concern … that fixes socially desirable, empowered space in metropolitan England or Europe and connects it by design, motive, and development to distant or peripheral worlds … conceived of as desirable but subordinate."[34]

Like Pratt's notion of "anti-conquest," Aguirre's formulation proves valuable insofar as it enables an analysis of self-avowedly non- or even anti-imperialist texts that remains attentive to the ways in which those texts and their authors nonetheless participated in an enterprise that can meaningfully be described as imperial. Such an analysis will be a frequent preoccupation of mine in the pages that follow. Nonetheless, Pratt, Burnett, and Aguirre all minimize what is to my mind one of the most noteworthy aspects of the Victorian encounter with Latin America – or, more exactly, of some of the various and diverse encounters the naturalists I study had with South American flora and fauna, including its peoples. For all of them, such encounters shake their sense of themselves, implicate them, change them. Rarely masters of all they survey gazing down on marvelous possessions with imperial eyes, these travelers more often find themselves enmeshed in and overawed by their surroundings. Writing about eighteenth- and early nineteenth-century scientific voyages of discovery in the South Seas, Jonathan Lamb contends that to "insist that the point of such voyages was exclusively the production of truth for imperial purposes is to ignore … the confusing and sometimes inexpressible experience of extreme conditions endured by the voyagers themselves."[35] More recently, adding to his focus on the South Seas an exploration of the experience of Latin American settlement, Lamb has described "settler metamorphoses" in which "there is no effectual loyalty to home, or any ground of a previous identity on which to anchor alterations of the Self, and make them obedient to will and intention."[36] In much the same way, what the travelers I study encounter as well as the conclusions they immediately or eventually draw disallow the sense of separateness and solidity necessary for easily held convictions as to their difference, superiority, or right to rule. In some instances, as in the case of Alfred Russel Wallace, those conclusions eventuate in an explicit and thoroughgoing anti-imperialist stance. In others, such as that of W. H. Hudson, sentiments that echo the young Huxley's with near exactitude are exposed as inimical to the best that the natural and human world contain.[37] The point here is not to extricate these naturalists from their involvement in imperialism; Victorian science and empire are inextricable, in their cases as in all others. It is, rather, to attempt to add further nuance to the increasingly sophisticated analysis of scientific discourse, informal imperialism, and narratives of encounter, and particularly to demonstrate that among the most perdurable elements of encounter was its ability to place verities in question.

One consequence of this attempt has been the decision to pay particular attention to those moments in the work of the naturalists I study where the self is overwhelmed by what it comes into contact with. Instructive in this regard is Darwin, who, writing in *The Journal of Researches* (1839), the book that in later editions would be known as the *Voyage of the Beagle*, notes of his first day in South America, February 29, 1832: "Delight itself … is a weak term to express the feelings of a naturalist who, for the first time, has been wandering by himself in a Brazilian forest … To a person fond of natural history, such a day as this, brings with it a deeper pleasure than he ever can hope again to experience."[38] Eliding his position as a member of HMS *Beagle*'s cartographic expedition, Darwin plunges at once into an affective and aestheticizing register. The result is an account of the naturalist as a creature of insistent desires – the gratification of which leads to a pleasure so overmastering that it cannot possibly be repeated. Confronted with an unfamiliar but longed-for scene, Darwin understands his feelings to be key, and they have to do first and foremost with "delight." The production of knowledge, the ostensible purpose of his travels, makes an appearance, but only as the implied point of convergence of satisfied desire ("pleasure"), science as affective undertaking ("the feelings of a naturalist"), and novelty ("for the first time").[39] Recapitulating a familiar Romantic narrative, he stages an encounter between a solitary wanderer and a natural world brimming with ineffable significance. The figure that takes shape in this autobiographical moment is so completely wrapped in and rapt by the enjoyment to be derived from a novel landscape that its old self is pried open, its former contours put under pressure. Versions of the same figure recur in the work of all the writers I study and, whether rejected in the name of established truths or embraced as the harbinger of new ones, they prove at once the most memorable and the most productive effects of their travels. For Darwin and Wallace in particular, feelings of shock and wonder at indigenous South Americans, recollected in tranquility, lead by tortuous and very different paths to their initially identical conclusions as to the fact of evolution as well as to their eventually opposed surmises about the implications of evolution for humankind.

Novelty features prominently in many such moments of disequilibrium, as it does in Darwin's musings on landfall in South America. The "feelings" about which Darwin writes are specifically those "of a naturalist who, *for the first time*, has wandered by himself in a Brazilian forest." He and others document an encounter understood and valued as new. At the same time, however, their excited discovery of a "new world" was always and constitutively the rediscovery of an old one. To begin with, the initial European exploration of the region in the fifteenth and sixteenth centuries was a familiar story. Charles Kingsley well conveys this familiarity when he writes of the Americas on the first page of *At Last: A Christmas in the West Indies* (1871): "From childhood I had studied their Natural History, their charts, their Romances, and alas! their Tragedies; and now, at last, I was about to compare books with facts, and judge for myself of the reported wonders of the Earthly Paradise."[40] The Victorian engagement with South America was of necessity a rediscovery insofar as it was always a case of "compar[ing] books with facts" – of encountering in the flesh, as it were, that which was already well known on paper.

But we can speak of a rediscovery in a more strictly scientific sense as well. In addition to maintaining restrictions on trade, the Spanish and Portuguese rulers of South America before independence limited travel in the interests of science conducted by representatives of other European nations.[41] There were, nonetheless, a few celebrated expeditions. Among the most important of these was Charles de la Condamine's 1735 voyage to ascertain the shape of the globe by calculating its dimensions at the equator, a voyage the successes and (mostly) catastrophes of which are amply documented in the wealth of texts that resulted. La Condamine as well as, later, the Prussian Prince Adalbert, Johann Baptist von Spix, and Karl Friedrich Philipp von Martius served as important predecessors for Victorian natural historians who traveled to South America.[42] For the Victorians themselves, however, it was above all Alexander von Humboldt who prepared the way for those who were to follow. Humboldt's account of his and Aimé Bonpland's 1799–1804 expedition, which eventually filled more than thirty volumes, was most readily available to Britons in the form of Helen Maria Williams's translation of the *Personal Narrative of Travels to the Equinoctial Regions of the New Continent* (9 volumes, 1814–29). The book christened the continent "new" again, in the same moment and by virtue of that rechristening ensuring it would be "old" (familiar, known, recognizable) to all those who came after. Humboldt definitively established the terms for how South America, and especially South American nature, was to be viewed, and praise of Humboldt's descriptions accompanied by disclaimers about their own inability to live up to them is a feature of the work of later writers so common as to constitute a veritable cliché.[43] Darwin may again be taken as representative:

As the force of impressions generally depends upon preconceived ideas, I may add, that all mine were taken from the vivid descriptions in the *Personal Narrative* of Humboldt, which far exceed in merit any thing I have read on the subject. Yet with these high-wrought ideas, my feelings were far from partaking of any tinge of disappointment on first landing on the shores of Brazil.[44]

As this pronouncement attests, so powerful was Humboldt's precedent that it obtained even for those parts of the continent, such as Brazil, where he never set foot. Of course, Darwin himself would come to occupy a similar position; after the publication of the *Journal of Researches*, his name joined Humboldt's as an inescapable point of reference and citation.

Victorian scientific travel and natural history in South America, then, were from the outset enterprises and discourses of belatedness. Accordingly, Victorian naturalists' relations to what they encountered were structured by a double discourse of loss and desire. Darwin's references to Humboldt, Wallace's and Hudson's to Darwin, Kingsley's to Sir Walter Ralegh and Sir Francis Drake – these and many other ritualized acknowledgements of precedence amount to a complex form of homosociality in which male naturalists place themselves in history by reenacting the travels of men who forestalled them. Science and exploration are of course self-consciously cumulative endeavors, so in one sense there is nothing unusual about such reenactment. But the insistence and intensity with which belatedness informs the work of these writers exceeds the bounds of what we might, recalling Thomas Kuhn, term "normal science" – with curious and far-reaching effects.[45] On one hand, they remain latecomers, their discoveries derivative, their texts copies rather than originals. On the other, in a curious valorization of belatedness, they refashion the proverbial significance of imitation by demonstrating that repetition is the sincerest form of memorialization.[46] They make their own exploits memorable by quoting remembered texts, tracing remembered routes, and recognizing remembered landmarks, hoping to secure in the process their own lasting fame as well as to extend that of those they repeat.

Describing the relationship between Robert Schomburgk, who was commissioned by the Royal Geographical Society to map British Guiana in the 1830s and 1840s, and his illustrious predecessor Sir Walter Ralegh, Graham Burnett employs the term "metalepsis," which in this usage signifies "invocations of history that are deployed in order to authorize even as they are stripped of their authority and content." "Rather than inaugurating imperial legitimacy in a single, tenuous (temporal or spatial) point," Burnett avers, "the circular structure of metaleptic cycles closed the authority of the past and present into mutually reinforcing hoops to bind territorial claims."[47] The concept is a powerful one that has been useful to me in clarifying what is at stake in, for instance, Hudson's foregrounding of his relation to Darwin. At the same time, it risks overestimating the degree to which the present can overwrite the past – the degree to which, that is, "invocations of history" are in fact subject to being "stripped of their authority and content." For the figures I study, the past, even if summoned to authorize their present, persistently threatens to render them echoes or afterthoughts.

The South America that Victorian naturalists traversed was thus at once "new" and "old," open to exploration but already traveled and already seen, and the negotiation of that double temporality informs their ability to dream of imperial futures for the region as well as to imagine the origins of the human in what they understood to be its savage past and present. Although South America almost entirely lacked a British imperial presence on the ground in the nineteenth century, the continent could still be seen as important to Victorian Britain's identity as an imperial nation-state by virtue of its pivotal significance in the sixteenth and seventeenth centuries, at the inception of Britain's maritime prowess and territorial expansion. And although South America lacked, too, literal proto-humans, and even the "man-like apes" to be found in Africa and Asia, its "savages" could be invoked in remembrance of the prehistory of so-called civilized humanity.[48] Imperial imaginings and remembered human origins come together in yet another kind of work structured by the negotiation of the familiar and the unfamiliar, the old and the new: natural history.

COLLECTION AS RECOLLECTION

Addressing the wide scope of the essays in their and James A. Secord's landmark volume, *Cultures of Natural History*, Nicholas Jardine and Emma Spary stress "how various are the frameworks that have structured and informed natural historians' dealings with nature, [and] how the boundaries between the natural and the conventional, artificial and social have been continually contested and relocated."[49] If one constant remains amidst such variety, contestation, and relocation, it is the place of the collection. Understood as a practice, natural history is above all a discipline of collecting. Thus Janet Browne observes that in the nineteenth century

[n]atural history, with its emphasis on the physical objects of nature, fitted the contemporary ethos as no other science could hope to do. Whether dealt with by experts or amateurs, the subject was almost entirely based on tangible things – on specimens of rocks and plants, on stuffed animals, drawers of minerals, serried ranks of butterflies and moths, jars of marine invertebrates, fern cases, aquariums, crates of bones.[50]

Some sense of both the importance Victorian naturalists placed on making collections and the staggering numbers involved is provided by Henry Walter Bates, Wallace's traveling companion in the Amazon for two years, the discoverer of a type of mimicry now called Batesian, and, as Assistant Secretary of the Royal Geographical Society from 1864 to 1892, an influential scientific administrator.[51] In a small notebook carried with him during his eleven years in the Amazon, 1848–1859, Bates kept a running tally of "Consignments of Collections to England." The first entry is dated August 24, 1848, almost exactly three months after his and Wallace's arrival, and begins:

Lepidoptera	1936
Coleop[tera]	776
Hymenop[tera]	409

The list continues, adducing the numbers for several other orders to this accounting of butterflies, beetles, and ants, resulting at the bottom of the page in a total of "3635 Insects" – an average of about forty specimens collected per day from the moment Bates stepped onto Brazilian soil to the moment of the reckoning.[52] Alex Shoumatoff reports the final tally of over a decade's work in the number of species rather than individual specimens: "14,712 species, no fewer than eight thousand of which were new to science."[53] Bates shipped these insects to his agent, Samuel Stevens, for sale to institutional and private collectors in Europe.[54] Such shipments were indispensable insofar as they provided the funds necessary for him to remain in South America. Further, they exemplified what being a natural historian was all about: capturing, describing, identifying, and carrying or sending away bits and pieces of the natural world.[55]

But collecting for Victorian naturalists was always also recollecting.[56] Ceaselessly on the lookout for new species, valuable in their own right as additional data about the natural world as well as for the possibility that their names might memorialize their discoverer, Victorian naturalists just as regularly recognized and collected examples of familiar species. (Eight thousand of Bates's Amazon species were new – but nearly seven thousand more were not.) Having encountered descriptions in scientific journals or travel narratives and, often, having studied actual specimens in museums or private collections, naturalists were in the position, to quote Kingsley once more, of "compar[ing] books with facts" – or, more precisely, of comparing remembered details of the morphology, coloration, anatomy, and so forth of described or preserved specimens with living exemplars found in the field. When recounting his first sight of the giant water lily *Victoria regia* in its native habitat, for instance, Wallace observes: "The leaf was about four feet in diameter, and I was much pleased at length to see this celebrated plant; but as it has now become comparatively common in England, it is not necessary for me to describe it."[57] Wallace experiences pleasure at seeing an exotic plant that is, nonetheless, already so familiar to him and his readers that description is superfluous. Belatedness, discovery as rediscovery, recognition of a novelty already known: these characterized not only the encounter with South American flora and fauna but also with the continent in its entirety.

The natural historical collection bears connection to recollection in another sense, too, for collected specimens, together with the notebooks, journals, sketches, and other records of a voyage, functioned as a kind of prosthetic memory that promised to recall times and places that might otherwise be lost.[58] Ann C. Colley nicely captures this aspect of collecting when, writing about Darwin, she compares his labeling of specimens to "laying steppingstones across the River Lethe."[59] Discovery, so often mythologized as the work of an instant, in the case of natural history played itself out over a lengthy period and involved prospect as well as retrospect, fieldwork as well as cabinet work.[60] Accordingly, the loss of collections, as I discuss in my chapter on Wallace, was the catastrophe of catastrophes.

Recollection also played a privileged role in terms of how naturalists wrote up their findings. At once remarkable and typical is the conclusion of Bates's own *The Naturalist on the River Amazons* (1863):

The saddest hours I ever recollect to have spent were those of the succeeding night [the night of his final departure from Pará, Brazil – now Belém] when, the mameluco pilot having left us free of the shoals and out of sight of land though within the mouth of the river at anchor waiting for the wind, I felt that the last link which connected me with the land of so many pleasing recollections was broken.[61]

Looking back from the present of the writing, Bates describes a liminal moment: "waiting for the wind" that will take him from Brazil to England, from past to future, he floats as if blind, still in the Amazon delta but unable to discern the shore. Remembered about that moment is not only his unhappiness at the prospect of leaving South America for the last time but also South America itself as a place of memories. Accompanied and intensified by a doubling of affect ("saddest hours," "pleasing recollections"), the double recollection at the close of *The Naturalist on the River Amazons* memorializes the moment of Bates's separation from the Amazon as if it were a separation from memory itself.

Natural history was a practice of recollection, then, because specimens encountered and collected – even quite rare and strange ones, as Wallace's comments on the *Victoria regia* attest – were often in the event re-encountered and recollected. It was also a practice of recollection insofar as natural-historical textual production characteristically borrowed the narrative form of the retrospect in order to frame travels as the events of, say, a novel are framed: as a potentially inchoate whole made coherent by the shared backward glance of reader and narrator. But there is still another sense in which one can speak of natural history in connection with recollection, one that can be discerned in Bates's vision of the natural world itself and South America in particular as a scene or site of memory.

A similar but fuller and more explicit vision is to be found in Grant Allen's "Introduction" to Rodway's *In the Guiana Forest*, where South American nature takes the form of what might be thought of as living memory or externalized recollection. For Allen, all tropical countries hold a special fascination because, as he puts it, "The tropics *are* nature."[62] To explain this apparently extravagant claim in which one part of the natural world stands as representative of the whole, Allen invokes struggle, diversity, and time:

It is not merely that there [in the tropics], and there alone, do you see life at its fullest, its fiercest, and its fieriest. It is not merely that there do you find the struggle for existence carried on with a wild energy which none can overlook, both among plants and animals. It is not merely that there trees, shrubs, and herbs, beasts, birds, and reptiles abound, with a richness and a variety unknown in more temperate regions. The tropics have a far deeper value for the biologist than all that. They are typical and central. They are ancient and historical. They represent in our cooled and degenerate world, from which all the most Titanic forms have fled, the circumstances under which plant and animal life first arose, and under which it passed through the main stages of its evolutionary history. This it is that gives value to the tangled brake of tropical scenery in the eyes of the biologist; he sees in that crowded and bustling woodland the image of the great world where our forefathers were nurtured.[63]

At the outset the tropics function as a synecdoche for all living things insofar as they exemplify nature as an arena of contest and change, featuring in its "fullest," "fiercest," and "fieriest" manifestation that struggle made famous by its place in the subtitle of Darwin's *On the Origin of Species by Means of Natural Selection, or The Preservation of Favoured Races in the Struggle for Life* (1859). Allen, an ardent Darwinian (the "tangled brake" he mentions is evidently an allusion to the famous "entangled bank" passage at the end of the *Origin*), elevates to representative status that part of the natural world he believes most clearly illustrates the forces of evolution at work.[64] Further, because they feature "a richness and a variety" of plants and animals "unknown in more temperate regions," the tropics provide, too, a compelling example of evolution's unbridled productivity. Two kinds of intensity required by evolutionary theory, intense struggle and intense diversity, account in large measure for Allen's sense of tropical nature as the instantiation of nature as such.

More oddly and powerfully, however, the interest of the tropics, their "far deeper value" because of their status as "typical and central," has to do with time. If the present state of the planet appears "cooled and degenerate" in relation to some primeval era peopled with "Titanic forms," the "ancient and historical" tropics preserve the key features of such an era intact. They preserve, that is, "the circumstances under which plant and animal life first arose, and under which it passed through the main stages of its evolutionary history." The tropics *are* nature, then, not simply because they offer the spectacle of intense struggle and diversity but also because, on Allen's reading of the history of the planet, that intensity represents a remnant of the past that has survived into the present. Indubitably part of "our cooled and degenerate world," the present-day tropics nonetheless embody and so provide access to a time before that world came to be. And while the promise held out by such access to the distant evolutionary past at first seems of narrowly scientific interest – Allen writes of the value of the tropics "for the biologist" and again "in the eyes of the biologist" – by the end of the passage the possibility of a pertinence at once more general and more immediate appears: in the tropics can be discerned "the image of the great world where our forefathers were nurtured." Relic and reliquary at once, the tropics enable us to gaze at the scene of our own origins.

It is in the context of this valuation of tropical nature as a whole that the declaration with which I began this introduction – "No tropical country is more interesting than South America" – must be understood. Like other tropical countries, South America for Allen provides a window onto earlier moments in evolutionary history, a picture of the origins of life and specifically of the origins of the human. But South America is distinct, the most interesting of tropical countries, because it "preserves, as in amber, numberless *intermediate* stages" of biological development.[65] Neither "very antique Australia" nor "very modern Asia," South America constitutes "a half-way house in the history of evolution, and possesses for many of us, therefore, that indefinite charm which the Middle Ages possess as a half-way house between Graeco-Roman civilisation and the squalid modern industrial system."[66] In this mapping of one tripartite historicity onto another, surely one of the most bizarre instances of Victorian medievalism on record, South America amounts to the Middle Ages of nature.[67] The metaphor adds complexity to the simple binary contrasting a primeval era of Titans with a sadly reduced contemporaneity. South America stands halfway; an "intermediate" stage between ancient and modern, its "indefinite charm" as well as its peculiar appropriateness as the site for remembering human origins inheres in its unfamiliar familiarity. Squalid modernity can perceive clearly in the South American tropics that which it has more difficulty recognizing in whatever the biological correlative of "Graeco-Roman civilisation" might be: the image of its own beginnings.

The difference between the tropics of the New World and the Middle Ages, however, is of course that the latter exist at an insuperable temporal distance from the present, approachable via artifacts such as the remains of buildings, roads, and weaponry, and also through texts, but only in fragments and only as that which has been irretrievably lost. South America, by contrast, although the past, is a present past, a living anachronism that functions for moderns such as Allen as a *lieu de mémoire*. Writing near the end of the nineteenth century, Allen deploys the trope of anachronistic space in connection with South American nature, contending that in that nature is to be found "some faint idea of that luxuriant world in which the fierce battle of the kinds was first fought out, and in which, as I at least believe, the early ancestors of man first began to be fairly human."[68] Now not precisely the past itself but rather its "faint idea" (its avatar, its trace, its commemoration), the South American tropics invite identification of the origins of the human in savagery ("the fierce battle of the kinds") and by way of memory.

SAVAGE MNEMONICS

In the texts I examine, several kinds of memory are at play at once. To begin with, personal recollection serves as a key textual modality. It is not just that most of these works were written up after the fact, from a point of vantage not only geographically but also temporally removed from the site of encounter. These writers deliberately cast their accounts as retrospects: looking back to an earlier period in their lives, they focus attention on, in, and through memory. Such individual or personal remembrance is bound up with memory of another sort. In the forests or on the plains of South America, nature itself is constructed and confronted as an externalized and collective form of memory, a remnant of and mnemonic for the heretofore inaccessible recesses of the biological past. Such constructions and confrontations frequently take place with other humans understood as themselves a kind of nature, anachronistic instances of early humanity. Remembering their personal pasts as well as the exploits of their predecessors, these writers remember, too, the past of all life, and particularly their own species past as humans. These multiple pasts collude and collide, marking moments of crisis as well as of consolidation.

Rodway's *In the Guiana Forest* itself provides a more detailed understanding of how the backward glance encoded in the very name natural *history* could function as a form of memory. With his subtitle, "Studies of Nature in Relation to the Struggle for Life," Rodway signals the interest he will take in the forests of South America as a site of evolutionary contest. Because of that interest, and because of his particular understanding of evolution as a matter of, in Herbert Spencer's coinage, the survival of the fittest, Rodway elaborates a description of the Guiana forest as a field of perpetual battle among innumerable individuals vying for light, space, and nutrients: "the fight is made up of single combats, where each forest giant is a centre with enemies in every direction."[69] The individuality attributed by way of personification to "each forest giant" (that is, each tree) results from Rodway's concern to portray the tropics as the scene of a sprawling, endless melee. Accomplishing that portrayal necessitates a view of every organic being as utterly solitary, cut off from the possibility of cooperation or community by the stringent requirements of the effort to endure. The tree-overshadow-tree world of the rainforest stands as an epitome of evolutionary nature as such.

This interpretation of Darwinism is familiar insofar as it approximates the individualistic, industrial-capitalist appropriations of evolutionary theory common in the late nineteenth century and beyond. Unexpectedly, however, Rodway's depiction of the South American forest also moves in a countervailing direction, one emphasizing connection and continuity rather than isolated individuality. Such connection is to be found, not among the various species struggling to survive in the present, but rather within a single organic entity reaching back into its own distant past. Rodway exhorts his readers to "look upon every individual man, beast, or plant as nothing less than the same being who lived tens of thousands (perhaps millions) of years ago, and has continued to live up to the present moment … We are our parents, grandparents, and ancestors of all past ages, up to that simple cell which first showed a germ of life."[70] Extrapolating from this claim, paired with the additional contention that the past of which each individual serves as a living embodiment is in some way available to that individual, Rodway arrives at the notion of a bodily or "physical" memory, a memory that he also refers to as "instinct." "It seems as if there are two memories," he explains, "one which permeates the whole body and belongs to the continuous line of generations, and the other that of the individual life. The former is necessarily the strongest."[71]

Several consequences follow from this positing of a dual memory. For Rodway, and in this he is more Lamarckian than Darwinian, evolutionary change amounts to the accumulated and transmitted effects of individual experience: "The experiences of every past generation are embodied in every living thing, and each one of these affects the offshoot more or less."[72] Further, what he calls the "theory of continuous existence" defines memory in such a way as to emphasize, not the recollection of those things that have occurred to a single entity in the course of its own lifetime, but rather something like the indelibility of entries in an ever-expanding notebook reaching back in time to the initial moment of

life on earth, the moment when "that simple cell" appeared "which first
showed a germ of life." "All these things go to prove that we have within
us," Rodway concludes, "a host of memories of which we know nothing
and that the record is being kept for all future generations."[73]

This use of the term "memory" to name marks left on or in the organism that pass from generation to generation is no longer current, but it
was a commonplace among the Victorians. Walter Bagehot establishes
just how uncontroversial such usage must have seemed when he employs
it as the basis for the entirety of his argument in *Physics and Politics* (1873):
"[T]he frame of each man [is] the result of a whole history of all his life,
of what he is and what makes him so, – of all his forefathers, of what they
were and of what made them so. Each nerve has a sort of memory of its
past life, is trained or not trained, dulled or quickened, as the case may
be."[74] As Laura Otis details in *Organic Memory: History and the Body in
the Late Nineteenth and Early Twentieth Centuries*, the concept of a corporeal memory that linked innumerable individuals vertically, through
the recesses of deep time, enjoyed a scientific vogue toward the end of
the nineteenth century. First formally propounded by Ewald Hering in
1870, among its most prominent English adherents was Samuel Butler,
whose vehemently anti-Darwinian *Unconscious Memory* (1880) seems to
lie behind Rodway's "physical memory."[75] Otis notes that such theories
of organic memory "placed the past *in* the individual, *in* the body, *in*
the nervous system; [they] pulled memory from the domain of the metaphysical into the domain of the physical with the intention of making it
knowable."[76] Memory in these theories names something corporeal and
unconscious, at once concrete and difficult of access. Literally within the
individual but neither of him nor easily brought to light (recalled, recollected) by him (and I use the masculine pronoun advisedly, for the subject
posited here is nearly always a male one), such a memory constitutes a
part of the individual's own past only insofar as that past is understood
as also, in fact perhaps as chiefly, the past of others: ancestors, precursors,
predecessors.

Gillian Beer notes the implications of Hering's and Butler's theories of
unconscious memory for thinking kinship between humans and other
animals as well as between "civilized" and "savage" humans: "Memory
thus becomes not a means of distancing the missing creature but the
motive-power for an identification with it. In such a re-definition, time
is not fixed and irreversible: memory can traverse heuristically the distances between the earliest moments and now, because the information
is conceived as continuous, embedded in the body and unconscious."[77]

Although widespread, such a notion of an organic or unconscious memory was not universally accepted in nineteenth-century scientific circles. Most Darwinians, Darwin among them, rejected it because it looked to the cumulative effects of experience rather than to natural or sexual selection for the mechanism of evolutionary change.[78] Even in Darwin, however, there is a sense that the body remembers what the mind of the individual cannot. In *The Descent of Man* he stresses that rudiments, vestigial organs, and "throwbacks" – individuals exhibiting atavistic characteristics – are incomprehensible except as evidence of a "community of descent" among "man and all other vertebrate animals."[79] Neither exclusively personal nor exactly a form of recollection, such corporeal manifestations of the past were nonetheless treated as memory – literally in Darwin's early writings, more figuratively later on.[80] This usage may seem strange given the current commonsense definition of "memory" as merely the capacity of individuals to recall earlier moments from their own lives. But it is our own usage that might well be viewed as the oddity, the result of the impoverishment of what was once a richer, more inclusive range of meanings. Nora's work on *lieux de mémoire*, among other things, de-individualizes memory, locating its crucial "sites" as social.[81] The memory posited by certain versions of evolutionary theory also imagines a transindividual mnemonic faculty: collective (recording "community of descent") and embodied (as much or more a matter of the *corpus* as of the mind).

In thinking through such conceptions and placing them in the context of the Victorians' engagement with memory more generally, I have found a measure of inspiration in Richard Terdiman's delineation of the fate of memory in nineteenth- and early twentieth-century Europe, *Present Past: Modernity and the Memory Crisis*.[82] Following Ferdinand Tönnies, Georg Lukács, and others, Terdiman identifies in the post-Napoleonic period the sense of an acute break between past and present, a break in the relation to history that manifests itself, Terdiman shows, as a break in and anxiety about the relation to memory.[83] Memory plays a contradictory role in this crisis: on one hand, in its formative and preservative aspect, it is credited with stabilizing identity, making of the past the matter of which the present and future are built; on the other, in its destructive aspect, memory and the past occupy the position of a "*problem* … a site and source of cultural disquiet."[84] "[W]ithin the atmosphere of such disruption," Terdiman observes, "the functioning of memory itself, the institution of memory and thereby of history, became critical preoccupations in the effort to think through what intellectuals were coming to call the 'modern.'"[85] As *Present Past* tells the tale, modernity came into being as a self-consciously distinct historical period by way of the distanciation of the present from the past as well as from its ability to recall or make sense of that past.[86] Modernity, on this representation, resembles an amnesiac who can remember only the necessity of recalling that which it might never have known or experienced in the first place.

Terdiman proves especially useful for his expansive definition of memory, which he glosses as "how the past persists into the present."[87] This definition is both capacious enough to admit the now-outmoded models of memory I call attention to and precise enough to frame the stakes of those models as having to do with the varied possibilities and failures of the past's continued existence. It is no accident that the stakes of evolutionary theory may be articulated in precisely the same terms. Further, unlike so many twentieth-century theorists and historians of memory, for whom the Second World War and the Holocaust inaugurate the West's mnemonic troubles, Terdiman dates the advent of the memory crisis to the beginning of the nineteenth century, thus placing the natural historians I study squarely in the midst of it. But here, too, lies the difficulty with my invocation of *Past Present* or indeed almost any other influential recent study of memory: science as such has little role to play.[88] The figures who loom largest in contemporary memory theory are poets, memoirists, novelists, sculptors, autobiographers, performance artists – in short, denizens of the "culture" half of the science/culture divide. But in this book I ask what place biologists and evolutionary theorists occupy in the upheavals Terdiman calls the memory crisis.

However unlikely it may seem, some indication of an answer to that question is provided by a thinker to whom so many adjudications of the relations between memory and modernity owe their basic assumptions: Walter Benjamin. For of course it is Benjamin who, in "On Some Motifs in Baudelaire," diagnoses the disappearance of memory as a result of the shock of the new in the form of industrialism, urbanization, and massification. The decline of ritual, the break-up of rural life and peasant culture, the ever-increasing reach of mnemonic technologies such as photography and the cinema: for Benjamin, these and other developments destroy memory as well as that relation to life memory once sustained, "experience."[89] Baudelaire figures as the writer in whose works evidence for these changes may be found, for he attempts to construct, writes Benjamin, "an experience which seeks to establish itself in crisis-proof form" even as he registers the impossibility of such a construction in the modern age.[90]

It is not often noted – and, if noted, not often made much of – that Benjamin's work, as he himself says of Henri Bergson's, "preserves links with empirical research. It is oriented toward biology."[91] Benjamin's diagnosis of modernity chronicles the impact of the crowd, new media, and haptic technologies on the human sensorium. As such, it might accurately be described as an account of the human animal in the throes of responding to dramatic changes in its environment. Moreover, "experience" itself seems to name a part of humans' biological heritage, for in an essay closely related to "On Some Motifs in Baudelaire," "The Storyteller," Benjamin suggestively links memory and experience to natural history: "Death is the sanction of everything that the storyteller can tell. He has borrowed his authority from death. In other words, it is natural history to which his stories refer back."[92] Biology, which stands as part of the explanatory framework for Benjamin's contentions about the effect of the modern, also provides him a way to articulate that to which the modern puts an end.

Following the lead of Benjamin's opaque hints about the relations between memory and biology, I posit natural history and evolutionary theory as key to the memory crisis. These intellectual formations and scientific practices admit of the same double description to which Terdiman gives modern memory. On one hand, evolutionary theory contributed to the alienation of the human present from its own past. Built on the assumptions of Charles Lyell's uniformitarian geology, in which the past of life on the planet receded into a dizzying abyss of time, it renders ultimate origins ungraspable. Lyell proclaims in *Principles of Geology*: "In vain do we aspire to assign limits to the works of creation in *space*, whether we examine the starry heavens, or that world of minute animalcules which is revealed to us by the microscope. We are prepared, therefore, to find that in *time* also, the confines of the universe lie beyond the reach of mortal ken."[93] Further, the more proximate beginnings deep time postulates for "civilized" humanity are animal and savage, beginnings widely resisted or recoiled from. When, for instance, in a legendary exchange, Samuel Wilberforce sought to confound T. H. Huxley by demanding to know whether it was from his grandmother's or his grandfather's side he traced his descent from apes, he counted on the shame and incongruity he assumed would attend the thought of such origins.[94]

But these very aspects of evolutionary theory also admit of a different, indeed opposite, characterization, one that emphasizes not separation from the past but rather connection to it. Insensible, gradual, continuous – the adjectives customarily invoked to explain evolutionary change posit a continuum between past and present, as did the effort to demonstrate the specificity of the descent of man. Thus in addition to the sort of disruption Terdiman and Benjamin claim for the modern, which problematizes the present's connection with the past, evolution also promises (and threatens) to link humans to the past as never before. In the travel narratives, scientific treatises, and autobiographies with which I am concerned, Victorian scientific travelers document their encounter with a South America they perceive at once as a realm of archaism from which they are estranged and a site that allows them, sometimes compels them, to remember and relive overlapping pasts.

   That they could imagine such remembering and reliving may be attributed in large part to the influence of recapitulation theory, succinctly conveyed by the dictum "ontogeny recapitulates phylogeny" – the surmise that, in the gloss provided by recapitulationism's most vociferous champion, Ernst Haeckel, "[t]he evolution of the germ (Ontogeny) is a compressed and shortened reproduction of the evolution of the tribe (Phylogeny)."[95] As the word "germ" indicates, strictly speaking the pertinence of the assertion has to do to with embryology: the development or "evolution" of an embryo, it was claimed, repeats in miniature the stages of evolutionary development the species to which it belongs has passed through. Although many historians of science view this so-called biogenetic law as a kind of heresy on Haeckel's part, Robert J. Richards makes the case that recapitulationism was fully congruent with the logic of nineteenth-century evolutionary theory and, moreover, that Darwin himself was a recapitulationist. According to Richards, Darwin "embraced the principle of recapitulation" because "it was essential to his theory that he do so."[96] Given the extreme imperfection of the geological record, the series of changes embryos undergo as they mature might provide the only empirical evidence on which a reconstruction of evolutionary history could be based. That recapitulationism appears so frequently in the work of the writers I study lends additional support to Richards's contentions. But it also reveals something more: the apparently irresistible slippage between embryological and more general or metaphorical versions of recapitulationism. For these writers, understanding themselves as subjects of evolution means being capable of repeating and, more staggeringly, revisiting in memory a range of pasts: their own, that of the travelers who came before them, that of Britain as an imperial nation-state, and that of humans conceived of as a species.

Taken together, those pasts give coherence to a human (modern, British) subject. They only do so, however, in and through faculties of recall that are, at best, equivocal. National memory, bodily memory, "racial" or species memory, and nature-memory enable the postulation of a subject understood relationally through time in a variety of ways: developmentally (the subject as culmination of a teleological biological or historical process); oppositionally (the subject as other to its own beginnings); spectrally (the subject as haunted by or an echo of earlier others, prior selves). But because these forms of memory, like simple personal reminiscence, can never promise absolute accuracy, the subject that emerges from their deployment can never achieve absolute coherence or solidity. The human constituted in these texts remains open to dissolution. Nevertheless, that human – a dominant version of the human as such, given the triumph of evolutionary theory – could not have been (cannot continue to be) imagined in the absence of the work of recollection I call savage mnemonics.

Chapter 1, "Charles Darwin's savage mnemonics," traces the consequences for the theory of evolution of Darwin's encounter with the inhabitants of Tierra del Fuego. More than forty years after that encounter, Darwin writes in his *Autobiography* (1887): "The sight of a naked savage in his native land is an event which can never be forgotten."[97] The truth of this observation in his own case is borne out fully, for at key moments in each of his major works on evolution by means of natural and sexual selection he recalls Fuegians and his reaction to them. Prosecuting the argument for human descent from other animals requires that Darwin find in Fuegians and other "savages" representatives of an otherwise empty space in the continuum of living forms, a missing link; accordingly, he does just that. Like South America itself for Allen, Fuegians occupy an intermediate position between the present and the distant past and so allow Darwin to remember and theorize the origins of humanity. But Darwin also recoils from the kinship with them that such theorizing mandates, a kinship he finds at once indubitable and insupportable. Too much like him to be disowned entirely but too alien to be accepted as kin (as well as, perhaps more tellingly, vice versa), Fuegians must be excluded from the very biological community their existence helps Darwin establish. Far from a curious footnote in the history of Darwinism, Darwin's recollections of Fuegians mark an exchange between remembering and forgetting, civilization and savagery integral to the development of evolutionary theory. They reveal the centrality of memory to the work of theorizing evolution and definitively inflect what kind of animal such theorizing imagines "man" to be.

Chapter 2, "Alfred Russel Wallace's tropical memorabilia," demonstrates that South American (and other) indigenes or "savages" were equally important in the thought and writings of the co-founder of evolutionary theory. Wallace's *A Narrative of Travels on the Amazon and Rio Negro* (1853), itself largely a work of memory in that fully half his notes were lost when the ship that was carrying him from South America back to England caught fire and sank in the Atlantic, chronicles a meeting with indigenous peoples that Wallace would repeatedly recall to the end of his long life. Recollecting that meeting and others like it initially afforded Wallace the ability to craft a theory independent of Darwin's but congruent with it in nearly every particular. But here the parallels between the two thinkers end, for while Darwin's memories of savages are held in a permanent state of tension or contradiction, a tension encoded in Darwinian evolutionary theory itself, Wallace's memories change over time – or, more precisely, their import does. For Wallace would eventually alter his views on evolution and argue that humans must be the product of factors beyond natural or sexual selection alone, an argument that draws its chief evidence from what he recalls about the physical perfection and extraordinary intelligence of "savage" peoples.

If Wallace, like Darwin, incorporates memories in and of South America to give shape to a certain kind of human, he marks a signal difference insofar as he does not recoil from but desires to remember and return to "savage" origins. Drawing out the romantic primitivism latent but refused in Darwin's version of savagery, Wallace reinflects the meaning of the human as the subject of evolution. Largely collapsing the temporal and developmental distance between "civilized" and "savage" humanity, he makes remembering the savage in the (supposedly modern) self at once more possible and less threatening an undertaking. Memory, and specifically memory of the savage, remains central, but the consequences of that memory are transformed. Among the most dramatic outcomes of such a transformation are Wallace's antagonistic stance to what goes by the name "civilization" as well as his opposition to European imperialism.

In the third chapter, "Charles Kingsley's recollected empire," I continue to pursue the question of imperialism. Here I do so in the case of an encounter with South America that never actually took place: the one envisaged by Charles Kingsley in *At Last: A Christmas in the West Indies*. Near the end of *At Last*, in a retrospective glance at what has been done and what left undone, Kingsley writes: "The hunger for travel had been aroused – above all for travel westward – and would not be satisfied. Up the Oroonoco [*sic*] we longed to go: but could not. To La Guayra and Caraccas [*sic*] we longed to go: but dared not."[98] Unable or unwilling actually to travel to South America, Kingsley can only gaze mournfully westward – and backward as well, to a lost past of mythic English imperial activity involving Sir Francis Drake, Sir Walter Ralegh, and other English "sea dogs," naturalists, and explorers. For Kingsley, the attractions of remembering and repeating outweigh those of discovering and pioneering. He desires not to best those precursors whose traces he finds everywhere but simply to follow in their wake. Embracing belatedness, Kingsley joys in being a latecomer to a place troped as the land of his own childhood as well as the childhood of his nation.

Kingsley's work formalizes and heightens the slippage between personal and biological memory evident in Darwin and Wallace. Further, it fuses biological memory with empire and nation in a more thoroughgoing manner. Whereas for Darwin "savages" were what people such as himself once were, and for Wallace people very much like what he and other Europeans now are (or could perhaps aspire to become), for Kingsley savagery is a state that, like the South American continent itself, may be longed for but can never be arrived at. The Kingsleyan subject is thus a subject postponed, suspended between memories to which he cannot return and desires (of settling down in the Americas, for instance) whose accomplishment would necessitate an impossible forgetting. By the same token, the Kingsleyan empire – in this resembling the empire envisaged by a number of influential theorists of and propagandists for "Greater Britain" at the time – can only exist as a once-and-future empire, a haunted thing made up of the past glory and future possibility that stand as bookends to its empty present.

In the fourth chapter, "W. H. Hudson's Memory of Loss," biology and personal recollection, empire and savagery, past and future are brought into their closest, most unnerving proximity. If Kingsley encounters the Americas as if he were encountering memories, for Hudson, born in South America but pursuing a literary career in London, writing about that continent was always writing about and with the aid of literal recollection. Further, as signaled by the titles of works such as the autobiography *Far Away and Long Ago* (1918) and the novel *The Purple Land that England Lost* (1885), those memories appear to take the familiar form of nostalgia. But such nostalgia is complicated insofar as the return it seems to long for is a double one: to Hudson's own childhood as well as to the childhood or savage beginnings of the human species. His books of essays on South American flora and fauna read alternately as autobiography, natural history, and ethnography. Exploring the implications of human descent from nonhuman animals and proto-humans, Hudson focuses specifically on the senses (his own and those of "savages") to argue for the continuity of savagery and civilization as well as the possibility of a two-way traffic between them. At the same time as he welcomes such a possibility, he fears it has already been or may soon be for ever foreclosed. Such fears reveal that the *lieu de mémoire* or site of memory that was South America for the Victorians was always a *lieu d'oubli* as well, a site of forgetting that threatened to sever the connection between modernity and the entire range of its savage pasts.[99] They also signal Hudson's insight that the doctrine of evolution posits an especially ineluctable forgetfulness: a leave-taking without return, a complete and final disappearance. Darwin exhorts us to remember the destruction omnipresent in the natural world but longs to forget Fuegians; Hudson fears that such forgetting is not only possible but nearly inevitable, taking with it all he values. By way of response, he appeals to the only "savage" safe from extinction, the one he imagines to reside in each "modern" human, in the service of a restoration (in fact, a creation) of a public, collective memory of wildness and savagery harnessed to political ends.

In a brief "Coda," I return to the twentieth- and twenty-first-century histories and theories of memory I have discussed at some length in this introduction to argue that, despite contentions to the contrary by writers ranging from Fredric Jameson to Andreas Huyssen, we do not yet live in a world given over to forgetting. Memory persists, even the forms of memory examined at length in this book. And while the continued presence of such savage mnemonics may seem atavistic, an embarrassing leftover from more barbaric days, I argue instead for the potential critical value of its insistence on memory as more than individual, and specifically as pertaining to the human conceived of as a species.

In a 1906 essay titled "Ancestral Memory: A Suggestion," Forbes Phillips illustrates the suggestion of his title with the following anecdote:

> As I walk along a dark lonely road, my ears are on the alert, I glance to right and left, I look over my shoulder. Where did I learn this habit? May it not be the memory-disc giving off its record? My savage ancestor learned by long years of experience to be specially on his guard in a lonely place, and in the dark.[100]

In this short passage are to be found three key strands of my argument, strands pursued in each of the chapters outlined above. First, Phillips posits an expansive definition of memory – so expansive as to include his own habitually suspicious actions while "walk[ing] along a dark lonely road." Second, there is the sense (literalized in the phrase "memory-disc") of such a memory as a technology, and more exactly as a technology of the human.[101] Third, the moment to which that technology allows access is a moment of savagery, and of savage ancestry in particular. At the complex point of suture where the three strands come together, Phillips comes into existence as a certain kind of subject. In this book, I locate the nodal moments in a genealogy of the possibility of such a suturing and suggest the degree to which it and the subject that resulted characterize modernity. By way of their experiences with and memories of South American "savages" and South America itself viewed as a place of savagery, I argue, Victorian natural historians worked to constitute "the human" in its modern form. Bringing a body of theory having to do with memory and science to bear on a rich archive of familiar as well as understudied texts, *Darwin and the Memory of the Human* aspires to reshape our understanding of the history of memory, to draw out the implications for studies of imperialism and transnationalism of the Victorian engagement with South America, and to examine the consequences of evolutionary theory – above all those consequences having to do with the definition of the "human" and its vicissitudes.

CHAPTER I

Charles Darwin's savage mnemonics

In *Idle Days in Patagonia* (1893), a remarkable nostalgic reverie about a journey to the land of its title, W. H. Hudson claims that what cannot be forgotten is that which moves the emotions most powerfully – in his case, the plains of Patagonia. Hudson further contends that the desolate landscape's memorability derives from its function as an instigation to remembering, and so returning to, an earlier mode of being. Describing the feelings that came over him during solitary rides on horseback through the Patagonian countryside, he writes:

my mind had suddenly transformed itself from a thinking machine into a machine for some other unknown purpose. To think was like setting in motion a noisy engine in my brain; and there was something there which bade me be still, and I was forced to obey. My state was one of *suspense* and *watchfulness*: yet I had no expectation of meeting with an adventure, and felt as free from apprehension as I feel now when sitting in a room in London.[1]

The cessation of thought, its replacement by apparently more primitive mental states such as suspense and watchfulness, the advent of an expectant orientation directed toward the future that nonetheless remains in an eternal present: for Hudson these changes indicate nothing other than a reversion to savagery. "[F]or I had undoubtedly *gone back*," he concludes:

and that state of intense watchfulness, or alertness rather, with suspension of the higher intellectual faculties, represented the emotional state of the pure savage ... If the plains of Patagonia affect a person in this way, even in a much less degree than in my case, it is not strange that they impress themselves so vividly on the mind, and remain fresh in memory, and return frequently; while other scenery, however grand or beautiful, fades gradually away, and is at last forgotten.[2]

Leaving civilization behind, the wanderer in an alien land is restored, for a time, to an elemental self, and so remembers ever after where and when that restoration took place: the sequence derives from a more or less standard version of "going native," a narrative in which becoming savage amounts to revisiting the prehistory of the human race, turning back the evolutionary clock so as to enable the return to an earlier (and sometimes, as in this case, better) mode of being. Such a narrative typically depends on the infectious presence of "actual" savages, and elsewhere in his work Hudson does recount finding in the aboriginal inhabitants of South America remainders and reminders of an earlier age of human development, revenant ancestors incongruously leading outmoded lives contemporaneous with their up-to-date descendants. Most notably, in the last chapter of *The Naturalist in La Plata* (1892) he reports finding a "living human [who] recalls a type of the past," a past so distant that "the volume of [its] history is missing from the geological record."[3] In a gesture Johannes Fabian diagnoses as the "denial of coevalness" and Anne McClintock labels the deployment of "anachronistic space," he portrays himself as occupying a present understood as the era of the modern while representing indigenous peoples or "savages" as out of step with modernity's forward march: in McClintock's words, "atavistic, irrational, bereft of human agency – the living embodiment of the archaic 'primitive.'"[4]

The stark account offered up in the chapter titled "The Plains of Patagonia," however, features only Hudson and the landscape. Going native in this instance does not entail (re)joining the reified memories that are present-day savages so much as turning the sensory deprivation of Patagonia to account in the service of remembering the savage within. The narrative is stripped bare, and this is possible only because of Hudson's belief that he himself has passed through a more primitive stage of development, a stage that has been preserved intact and to which he can thus return. Implicitly endorsing a metaphorical version of Ernst Haeckel's dictum that ontogeny recapitulates phylogeny, which is to say that the embryological development of the individual organism repeats in condensed form the entire series of evolutionary phases that the species to which it belongs has passed through, Hudson further imagines that certain moments along the phylogenetic path to the present can be recalled and revisited.[5] He believes, in short, not simply that he can remember being a savage but that, via memory, he can become one again.

In viewing nineteenth-century Europeans and European Americans such as himself as the end point on an evolutionary time-line, a culmination in relation to which other peoples and cultures represent so many fascinating or disturbing but at any rate defunct stops along the way, Hudson stands heir to a long line of thinkers. As Robert J. C. Young notes, early in the nineteenth century it was John Stuart Mill who, borrowing from historiographical models developed by Scottish Enlightenment philosophers, "formalized [the relation between civilized and savage] as a hierarchy of the historical stages of man, bringing geography and history together in a generalized scheme of European superiority that identified civilization with race."[6] Significant contributors to such developmental or stadialist thinking after Mill include T. H. Huxley and E. B. Tylor, both of whom Hudson cites.[7] But for Hudson, and indeed for virtually everyone writing on or thinking about "savages" in the second half of the nineteenth century, the paramount point of reference is not Mill, Huxley, or Tylor but Charles Darwin. Hudson can remember being savage because he cannot forget Darwin. Darwinian theory persuasively anchored the trope of anachronistic space in biology, giving scientific credibility to what had been an essentially figurative assertion. More important, and more remarkably as well, Darwin's own textual production from first to last takes shape in relation to savagery: for Darwin at his desk in Kent no less than for Hudson on the plains of Patagonia, the savage is unforgettable.

Writing over a period from the 1830s to the 1870s, Darwin repeatedly invokes savages: in the midst of, and then again when bringing to a close, the account of his youthful circumnavigation of the globe; in the last few pages of the first published full-scale elaboration of the theory of evolution by means of natural selection; as part of the conclusion of his efforts to demonstrate the applicability of that theory to humans; and, finally, when drawing up a collection of autobiographical notes on his life near its end. Although he sees savages as lacking a sense of history and hence possessing no meaningful past of their own, they refuse to be relegated to Darwin's past.[8] Returning to them at crucial moments of introspection, retrospection, and summation, he appears to be as incapable of moving beyond them as he believes them to be of moving beyond their state of savagery without the European intervention that will either civilize them or lead to their extinction. While at points material to the theoretical explication and empirical demonstration of the workings of evolution – their supposed position between "civilized" humans and other animals makes them indispensable to the argument that, like all other organic beings, humans evolved from earlier forms of life – savages in their various appearances in Darwin's work exceed evidentiary and forensic functions. The shock he evinces at the sight of them and the thought of their proximity to "civilized" humanity ensures that he cannot forget them. That same shock, however, also demands that savages be, if not precisely forgotten, then at least displaced from their position in a theory to which they nevertheless remain ineluctably central.

In what follows I trace the workings in Darwin's corpus of this savage mnemonics, which comprises two distinct but conjoined elements. On one hand, Darwin exploits a mnemonics of savagery, turning to savages in remembrance of the origins of the civilized. In doing so he takes up and strengthens the hierarchical narrative of human development running from the Scottish Enlightenment through Mill to Hudson and beyond. On the other hand, Darwin refuses the anamnesis afforded by such a mnemonics. Postulating human evolution requires that he distance himself from the savage, that he try to forget or erase the savage's kinship with the civilized. This vacillation between total recall and deliberate amnesia may appear idiosyncratic or pathological – some Darwinian version of Freud's repetition compulsion, perhaps, with the trauma of initial contact repeated endlessly in a fruitless search for mastery. Its insistent presence, though, marks a signal moment in the Victorian relation to the primitive: a definitive version of the familiar story of civilization's construction of itself in opposition to savagery, such vacillation reveals the paradoxical imperative within that story for the civilized, in constituting themselves as such, to take on the "savage" attribute of disregarding the very past that makes them who they are. What I call savage mnemonics also clarifies the nature of the work of memory demanded by the effort to grasp evolution, revealing, in the context of evolutionary theory, the imperative to remember as a necessary precursor to and component of the ability to understand. Finally, insofar as it stands as an exception to and even inversion of that work, Darwin's savage mnemonics shows that, in a world otherwise completely assimilated to the natural and the animal, in which "man," as Nancy Stepan puts it, is "merged ... in animal nature," savages for Darwin occupy the sole remaining locus of the human.[9]

I

Darwin's initial and defining encounter with savagery took place in South America during the journey around the globe he completed aboard HMS *Beagle* between 1831 and 1836, a journey that might fairly be described as having had everything to do with time. Robert FitzRoy, captain of the *Beagle* during those years, writes that he and his crew set out on the voyage "entertaining the hope that a chain of meridian distances might be carried around the world."[10] To this end, the ship maintained a battery of no fewer than twenty-two chronometers. These were deemed so essential by FitzRoy that he paid for some of them out of his own pocket, for their careful use would establish precise longitudinal positions and aid in the successful completion of the *Beagle*'s primary charge from the Admiralty, the mapping of the southern coastline of South America. This mapping itself had another kind of temporal resonance. It was timely in that, in the wake of a series of independence struggles touched off by the Napoleonic invasion of the Iberian peninsula, the Americas in the 1830s were for the British once again a "new world": newly opened to trade, investment, and exploration after centuries of inaccessibility.[11] Darwin biographer Janet Browne notes: "The whole point of the British Admiralty's desire to chart southern Latin America was to enable informed decisions to be made on naval, military, and commercial operations along the unexplored coastline south of Buenos Aires and to enable Britain to establish strong footholds in these areas, so recently released from their commitment to trade only with Spain and Portugal."[12] And in still another temporal register, Darwin's own scrutiny of the geology, flora, and fauna of the region provided him with key data on which he would construct what might be termed a "new" or at least newly discovered past for life on the planet, that past implied by the theory of evolution by natural selection propounded in *On the Origin of Species* (1859). At first a taxonomic exercise in filling out more and more blank spaces in Linnaeus's static catalogue of life, Darwin's assiduous investigation of the living archive that was South America eventually contributed to the transformation of taxonomy into genealogy, opening up a biological time as profound as the "deep" geological time Charles Lyell proposed in his *Principles of Geology* (1830–33). "In vain do we aspire to assign limits to the works of creation in *space*," Lyell wrote, "whether we examine the starry heavens, or that world of minute animalcules which is revealed to us by the microscope. We are prepared, therefore, to find that in *time* also, the confines of the universe lie beyond the reach of mortal ken."[13]

Occupying a strikingly complex position in relation to such novel temporal depths are savages and Darwin's relations with and memory of them. To attempt to get at that complexity, fraught as it is with the question of looking back, it seems best to begin with Darwin's most evidently retrospective work, the posthumously published *Autobiography* (1887). Written over several months in 1876, six years before its author's death, the slim *Autobiography* resembles a desultory assortment of anecdotes much more than the masterful marshalling of great moments in the life of a great man that might be expected. Like more recognizable autobiographical enterprises, however, Darwin's, too, is placed under the sign of memory: the top of the first page bears the words "Recollections

of the Development of my mind and character."[14] Such a title assigns two
roles for memory in the work to follow. "Recollections" promises something resembling that deliberate use of memory peculiar to autobiographical production: the selection and arrangement of remembered moments
from childhood and youth so as to reconstruct a past sufficient to explain
the adulthood that follows as culmination. Because Darwin seeks to give
an account, though, not only of the events of his life but also of "the
Development of [his] mind and character," memory in the *Autobiography*
is not understood as a technique of composition alone, nor as a mere
repository of raw material; it is also an object of scrutiny in itself.

When, therefore, Darwin turns near the end of the text to an inventory
of the mental faculties that enabled him, as he puts it in characteristically
self-effacing fashion, to "[influence] to a considerable extent the beliefs of
scientific men on some important points," he addresses his own faculties
of recollection directly:

> My memory is extensive, yet hazy: it suffices to make me cautious by vaguely
> telling me that I have observed or read something opposed to the conclusion
> which I am drawing, or on the other hand in favor of it; and after a time I can
> generally recollect where to search for my authority. So poor in one sense is my
> memory, that I have never been able to remember for more than a few days a
> single date or line of poetry.[15]

Articulating with disarming candor a certain kind of scholarly recall,
this statement describes a memory that, "extensive" but at the same time
"hazy" or "poor," might be expected to pose difficulties for the production
of a text like *The Descent of Man, and Selection in Relation to Sex* (1871),
more than 800 pages in length and replete with minute behavioral and
morphological details about an immense variety of animals – including,
crucially and infamously, humans. To enable that production, to solidify
the extensiveness and minimize the haze, Darwin fashioned an elaborate
prosthesis for himself: elsewhere in the *Autobiography* he tells of the series
of abstracts and indices he made of each book he ever read, mnemotechnic and compositional aids that he consulted anew each time he sat down
to work.[16]

The employment of these elaborate memoranda, immensely suggestive
insofar as it can be traced back to FitzRoy's meticulous habits of note-
taking and log-writing, which Darwin observed and emulated while
aboard the *Beagle*, and hence can be shown to have immediate and intimate connections with British imperial naval protocol, is something to
which I will return.[17] For the moment, I would like to consider another

passage from the *Autobiography*, one focused on a memory that stands out in high relief: that of Darwin's first contact with what he elsewhere calls "an untamed savage."[18] Near the end of a section titled "Voyage of the 'Beagle': from Dec. 27, 1831 to Oct. 2, 1836," Darwin singles out the three most memorable aspects of that voyage. Pride of place goes to "[t]he glories of the vegetation of the Tropics"; listed first, those glories also "rise before [his] mind at the present time more vividly than anything else." Next, anticipating Hudson and preparing the way for his remembrances there, come "the great deserts of Patagonia," a sublime landscape that "has left an indelible impression on [Darwin's] mind." Third and last in this inventory of those things that a hazy memory has no trouble recollecting with clarity are savages: "The sight of a naked savage in his native land," Darwin writes, "is an event which can never be forgotten."[19] Although he leaves the specifics to be inferred, the reference is certainly to a Fuegian, one of those inhabitants of the southernmost tip of South America that he first encountered when the *Beagle* made landfall in Tierra del Fuego in December of 1832.

Featuring in the *Autobiography* as one of the episodes of the *Beagle's* journey that remains sharpest in his recollection when he sits down to write a retrospective account of his life, the encounter with Fuegians makes its initial public appearance in Darwin's work forty years earlier, in the text that in later editions would come to be known as *The Voyage of the Beagle*: the *Journal of Researches into the Natural History and Geology of the Various Countries Visited by H. M. S. Beagle*, Darwin's contribution to the four-volume *Narrative of the Surveying Voyages of His Majesty's Ships Adventure and Beagle* (1839). Like the passage from the *Autobiography* from which I have quoted above, the "retrospect" at the end of the *Journal of Researches* includes a catalogue of those scenes and things "deeply impressed on [Darwin's] mind" over the course of the journey. Prominent among them, as in the *Autobiography*, is to be found the following: "Of individual objects, perhaps no one is more certain to create astonishment than the first sight in his native haunt of a real barbarian, of man in his lowest and most savage state."[20] Astonishing at first sight and unforgettable ever after, savages or barbarians obviate the need for abstracts, indices, or extensive notes of any sort.[21] Fuegians make an indelible impression. They memorialize themselves.

The reasons for their memorability are not far to seek. In the chapter of the *Journal of Researches* in which Darwin narrates his first sight of Fuegians, he claims that "[i]t was without exception the most curious and interesting spectacle [he] had ever beheld" (*J* 178).[22] "[C]urious" and "interesting" seem to assimilate Fuegians to parrots, volcanoes, the fossilized remains of giant sloths, and other natural-historical wonders on offer in South America, and the word "spectacle" immediately bears out that connection. But the passage quickly leaves such intimations of gentlemanly spectatorship and cabinets of wonder behind, replacing them with the admission of a shock nothing less than categorical in its intensity. Fuegians figure not simply as curious and interesting spectacles but as barely human, preternatural, alien: "Viewing such men, one can hardly make oneself believe they are fellow-creatures, and inhabitants of the same world" (*J* 184). Naked or nearly so despite inhabiting a land constantly subject to powerful storms and frigid temperatures, using few tools and poorly made boats, rumored to practice cannibalism (and in particular to consume elderly women in times of famine), without identifiable leaders, apparently without even basic social organization – Fuegians for Darwin incarnate the nadir of existence. They lack not only the accomplishments of other humans but also that joy in life occasionally evinced, according to Darwin, by such non-human animals as birds and even ants. "These were the most abject and miserable creatures," he announces, "I any where beheld," and in a footnote to that statement ventures the judgement that "in this extreme part of South America, man exists in a lower state of improvement than in any other part of the world" (*J* 184, 184n).

In this estimate of Fuegians, Darwin follows tradition. When Captain James Cook landed in Tierra del Fuego in January of 1769, for instance, he found the inhabitants "perhaps as miserable a set of People as are this day upon Earth."[23] Peter Hulme observes of both Patagonia and Tasmania that because they were conceptualized as "the 'uttermost' parts of the earth, the farthest south that a European could travel … , their populations inevitably became cast as the 'lowest' … forms of life."[24] For Darwin, the appearance of such debased misery, disturbing in itself, is still more so in that it shows every sign of immunity to change – is essentially ossified, fossilized. He writes that what little skill Fuegians possess "may be compared to the instinct of animals" because "it is not improved by experience: the canoe, their most ingenious work, poor as it is, has remained the same, as we know from Drake, for the last two hundred and fifty years" (*J* 185). With no past such as Darwin has – their canoes remain as ineptly constructed as ever, whereas the mention of Sir Francis Drake alludes to a world that Darwin is both heir to and confident of having left behind – Fuegians fail even in the ability to imagine a time or place beyond the one they inhabit, a failing troped in Darwin's account by their inability to conceive of a hereafter: "Captain Fitz Roy [*sic*] could never ascertain that the Fuegians have any distinct belief in a future life" (*J* 215).

Abject, miserable, lacking a sense of history, incapable of progress and so shut out from futurity, unable to envision even the consolations of a life beyond death – as if all this were not enough, Fuegians are finally and, indeed, primarily memorable for Darwin because they are themselves instigations to memory, living mnemonic devices. So much, at any rate, is conveyed by the full passage in the "retrospect" at the end of the *Journal of Researches* from which I have taken excerpts above:

Of individual objects, perhaps no one is more certain to create astonishment than the first sight in his native haunt of a real barbarian, of man in his lowest and most savage state. One's mind hurries back over past centuries, and then asks, could our progenitors have been men like these? Men, whose very signs and expressions are less intelligible to us than those of the domesticated animals; men, who do not possess the instinct of those animals, nor yet appear to boast of human reason, or at least of arts consequent on that reason. (*J* 474)

This excursus occurs amid a careful weighing of "the advantages and disadvantages, the pains and pleasures" of a circumnavigation of the globe (*J* 471). The account of pains, such as lack of privacy, loss of contact with family and friends, and perpetual sea-sickness, is set in the balance against the myriad pleasures involved in seeing tropical scenery and visiting a variety of foreign lands, gazing down at coral reefs and up at the constellations of the southern hemisphere. But this deliberate and symmetrical calibration of good and bad is brought up short when the list passes from scenery to human inhabitants in general and Fuegians in particular. Astonishing rather than advantageous or disadvantageous, painful or pleasurable, Fuegians disrupt the organizational structure of the passage in which they appear, refusing to take their place in the litany. Such disruption, signaled by, among other things, the odd play of interrogation and declaration (the initial question seems to answer itself but is immediately followed by an answer that resembles still another question), is due to the fact that the memory of the first encounter with "man in his lowest and most savage state" is tied to and in some sense about a memory of another sort: "One's mind hurries back over past centuries, and then asks, could our progenitors have been men like these?" Fuegians, that is, disturb insofar as they give rise to something akin to racial memory in the observer.

Walter Benn Michaels notes that the "obvious objection" to the use of the term "memory" in this way, as a name for Darwin's reflections upon his own ancestry set off by the sight of Fuegians, is that "things we are said to remember are things that we did or experienced whereas things that are said to have taken place in the historical past tend to be things that were neither done nor experienced by us."[25] The assertion that Fuegians cause Darwin to recall having been a savage himself, then, depends on a slippage in meaning between the individual-biographical and the species-historical. In the notebooks he kept from 1836 to 1839, and particularly in the 1838 notebooks devoted to transmutation and the biological basis of human and animal behavior, Darwin's speculations on memory nearly all turn on such a slippage because it seemed to provide an explanation for instinct. Thus in the so-called M notebook, written between July and October 1838, he observes: "Now if memory <<of a tune & words>> can thus lie dormant, during a whole life time, quite unconsciously of it, surely memory from one generation to another, also without consciousness, as instincts are, is not so very wonderful."[26] As Laura Otis demonstrates, Darwin would later move away from this literalization of the association of memory with inheritance, contesting it in the work of Ewald Hering and other theorists of "organic memory."[27] My interest in this chapter is with Darwin's insistent invocation of memory not as an actual mechanism of inheritance but as a trope he elaborates in connection with Fuegians, a trope that at once troubles him and provides him with the ability to negotiate relations among Fuegians, himself, and nineteenth-century Europeans generally. For seeing them causes Darwin to indulge in the quintessentially Victorian speculation that he might be face to face, if not literally with his own ancestors, at least with people very like them.[28] And the thought is not a happy one, for even "domesticated animals" appear preferable to these inexplicable and backward creatures, hideous skeletons in the familial closet. It is, however, haunting: Darwin would never forget it.

2

Like and unlike, near yet far, still alive and well but also outmoded and destined to disappear: such is the familiar stuff of the Victorian imaginary when it comes to savages. In *Victorian Anthropology*, George Stocking identifies these seemingly contradictory assumptions as fundamental to the comparative method, which is to say "the idea that in the absence of traditional historical evidence, the earlier phases of civilization could be reconstructed by using data derived from the observation of peoples still living in earlier 'stages' of development."[29] The more widespread and popular version of this idea, still quite familiar, may be glossed with Peter

Mason's succinct formulation as the "elision of the primitive and the primeval."[30] But Darwin's account is different from and more interesting than most incarnations of this comparatism. For, first, his encounter with Fuegians takes up a privileged place in the memory of his own early adulthood, always featuring as a significant part of what the *Beagle* voyage meant to him. And, second, Fuegians and the "savages" for whom they stand end up as both precisely what he must remember and what he needs to attempt to forget in order to defend the theory of evolution by natural selection and then apply that theory to humans.

Darwin had, of course, been in contact with Fuegians before his arrival in Tierra del Fuego, simply not, to use his own language, "untamed" ones. Three Fuegians were aboard the *Beagle* on its December 27, 1831 departure from Devonport: two men, called York Minster and Jemmy Button by their English captors, and one woman, known as Fuegia Basket. The three had been seized (or, in the case of Jemmy Button, purchased – his name commemorating his price) and taken to England during an earlier voyage of the *Beagle* and its sister ship, the *Adventure*.[31] Having been taught some English and instructed in Christianity as well as in such niceties as table manners and European dress – Jemmy Button's dandyism was eventually lampooned in the British press – they were now to be returned to their homeland. Gillian Beer has argued that the time Darwin spent among these Anglicized Fuegians crucially shaped his response to the people he saw in South America late in 1832: "Darwin's encounter with Fuegians in their native place gave him a way of closing the gap between the human and other primates, a move necessary to the theories he was in the process of reaching. But it came after his experience of Fuegians abroad."[32] In the event, both sorts of Fuegian proved requisite to theorizing evolution. For to advance "savage" humans as part of the evidence for human descent from animals, Darwin required that Fuegians be brutish and abject, and so plausibly close to animals, but also educable, and so like the rest of humanity.[33]

Something of this doubleness is made explicit in the *Descent*, where Darwin observes: "The Fuegians rank amongst the lowest barbarians; but I was continually struck with surprise how closely the three natives on board H.M.S. 'Beagle,' who had lived some years in England and could talk a little English, resembled us in disposition and in most of our mental faculties" (*D* 1: 34). Eliding the distinction between "untamed" Fuegians and "the three natives" on the *Beagle*, Darwin here collapses the abject and the educable savage into a single collective entity – "[t]he Fuegians." But his more frequent practice is to treat "untamed" Fuegians as representatives of savages generally and to distinguish Button, Basket, and Minster sharply from their less well-traveled kin. He observes in the *Journal of Researches*, for instance, that "[i]t was interesting to watch the conduct of the savages, when we landed, towards Jemmy Button" (*J* 180).

Such moments of distinction feature an incoherence or uncertainty that may usefully be clarified by contrast with other, contemporaneous accounts of Fuegians. In *Proceedings of the First Expedition, 1826–1830, under the Command of Captain P. Parker King*, the first volume of the *Narrative of the Surveying Voyages of H.M.S. Adventure and Beagle*, Parker King notes of his own first sight of the inhabitants of Tierra del Fuego:

> they appeared to be a most miserable, squalid race, very inferior, in every respect, to the Patagonians … [Their] seeming indifference, and total want of curiosity, gave us no favourable opinion of their character as intellectual beings; indeed, they appeared to be very little removed from brutes; but our subsequent knowledge of them has convinced us that they are not usually deficient in intellect.[34]

Like Darwin, King makes sense of Fuegians by invoking a comparison: immediately measuring Fuegians against Patagonians, King characterizes the former by finding them lacking, "inferior in every respect." The effect of King's comparatism, however, is quite different from that of Darwin's, for King makes a place for Fuegians, including rather than excluding them. Whereas Darwin's surmise that Fuegians resemble "our progenitors" contributes to establishing their uniqueness, their position as incarnations of "man in his lowest and most savage state," marooned in the backwaters of time, King's comparison of Fuegians to "Patagonians" connects them to other actually existing humans and, in so doing, acknowledges their inhabitation of a present to which King himself also belongs. Moreover, King concludes with the mention of a later and different estimation, one that marks his first opinion as erroneous: "but our subsequent knowledge of them has convinced us that they are not usually deficient in intellect." Revisiting and revising his initial judgement of Fuegians at the end of the very sentence in which that judgement is given, King appears to fault himself for too hastily arriving at a conclusion that he later determines was mistaken.

At the beginning of the *Proceedings of the Second Expedition*, FitzRoy, describing his own response, revises King's conclusions about Fuegians still further:

> During the time which elapsed before we reached England, I had time to see much of my Fuegian companions; and daily became more interested about them as I attained a further acquaintance with their abilities and natural inclinations.

Far, very far indeed, were three of the number from deserving to be called savages – even at this early period of their residence among civilized people – though the other, named York Minster, was certainly a displeasing specimen of uncivilized human nature.[35]

Especially notable is the explicit rejection of the applicability of the term "savage" to Jemmy Button, Fuegia Basket, Boat Memory, and York Minster, a rejection FitzRoy accounts for not by invoking whatever bits and pieces of English culture were urged on them during their time in England but rather by putting into evidence what he considers to be "their abilities and natural inclinations." The close contact allowed by the Fuegians' presence on the voyage back to England after their capture both elicits interest and gives the lie to a rush to judgement that would find in them confirmations of preconceived notions about savagery. Even the final reference to York Minster, although pointedly characterizing him as "a displeasing specimen of uncivilized human nature," achieves that characterization by individualizing him and thus the others as well. Not "the Fuegians" but one among their number "deserves" the appellation "savage."

Addressing FitzRoy's and Darwin's divergent reactions to Fuegians, Nigel Leask contrasts the former's "Christian anthropology" with the latter's "darker, more deterministic reading of bio-geographical pressure."[36] Despite the pro-slavery stance that occasioned trouble between him and the staunchly abolitionist Darwin early in the *Beagle* voyage, FitzRoy was committed to the monogenist position that all humans derived from the same aboriginal stock. He thus approached Fuegians as a paternalist and philanthropist: for him they were improvable, endowed with the potential to be Christianized and civilized. Such a view may have produced results both farcical and tragic, as the subsequent history of the three prodigal Fuegians demonstrates, but at base it presumed that savages shared essential "abilities and natural inclinations" with "civilized" humans.[37] For FitzRoy, even York Minster possessed a "human nature," however "uncivilized."

At the moment of arrival in Tierra del Fuego, despite his reaction to Fuegians as if to aliens or demons, Darwin shared FitzRoy's Christian monogenism. As Janet Browne notes, he "had no foundation for believing that ... native Fuegians were in some way closer to animals than he was. It was this rejection of separate origins that made the whole experience of Tierra del Fuego so painfully interesting to him."[38] Subsequently, Darwinian theory would play a key role in the biologization of race that Nancy Stepan, George Stocking, and others have shown constitutes the principal contribution of nineteenth-century biological and anthropological conjecture to the question of the meaning of human difference.[39] But in connection with Darwin this tells only half the story – or, rather, tells the story as if its grounds were not in the process of being shifted. It also neglects those moments when Darwin himself reads the "savage" peoples he encounters in the way that King and FitzRoy read them: as no different in essence from the civilized; in FitzRoy's words, as "ignorant, though rather intelligent barbarians."[40]

Predictably enough, given the history of European fascination with them, on the *Beagle* voyage it is Tahitians who seem most promising to Darwin in this regard.[41] Of his visit to Tahiti he writes in the *Journal of Researches*: "I was pleased with nothing so much as with the inhabitants ... There is a mildness in the expression of their countenances, which at once banishes the idea of a savage; and an intelligence, which shows they are advancing in civilization" (*J* 376–77). Other native peoples encountered include Auracanians in southern Chile, Australian aborigines, and Maoris of New Zealand. But memorable above all others are Fuegians, in whom Darwin finds no "mildness" and scant signs of "intelligence" or advance in civilization. Indeed, his obsessive recourse to the moment of his land-fall in Tierra del Fuego evidences little but the negative half of what Marianna Torgovnick, in *Gone Primitive: Savage Intellects, Modern Lives*, calls Western modernity's "cherished series of dichotomies" about so-called savages. In Western representations, she points out, savages are "by turns gentle, in tune with nature, paradisal, ideal – or violent, in need of control; what we should emulate or, alternately, what we should fear; noble ... or cannibals."[42] Astonished to the point of horror on first viewing Fuegians, incapable of forgetting them ever after, Darwin's repeated references to Fuegians attempt to place distance between their perceived barbarity and himself, his (sympathetic) readers, and his work – even as those references serve as indispensable evidence of evolution. The epitome of savagery, so significant for and troubling to Darwin are Fuegians that he invokes them in the concluding pages of the two central texts of Darwinian evolutionary theory, *On the Origin of Species* and *The Descent of Man, and Selection in Relation to Sex*.

3

In the *Origin*, Darwin scrupulously avoids speculation on the applicability of the theory of evolution to humans. The subject is touched on near the end of the text, and then only in three coy sentences: "In the distant future I see open fields for far more important researches. Psychology will be based on a new foundation, that of the necessary acquirement of each mental power and capacity by gradation. Light will be thrown on the origin of man and his history."[43] Concerned to establish the evolution by natural selection of plants and (non-human) animals, he prosecutes his argument with painstaking attention to the geographical distribution, rudimentary organs, color patterns, nesting habits, and so forth of pigeons, cats, ants, ostriches, geese, bees, beetles – an entire menagerie. Savages show up rarely, and then almost exclusively in connection with animals. Australian aborigines are mentioned, for instance, with an eye toward the dogs they keep and the livestock they do not (O 215). Or again, the impossibility that a generic "savage" could know in advance whether certain species are inherently more variable than others gives the lie to the assumption "that man has chosen for domestication animals and plants having an extraordinary tendency to vary" – and thus helps to establish the variability of all living things requisite to providing the raw material on which natural selection operates (O 17). In the "Conclusion," however, which recapitulates the book's "one long argument" (O 459) and then speculates on some of its consequences, savages qua savages put in a brief and singular appearance as those beings who epitomize the attitude to nature that must be rejected if evolution is to be understood and appreciated:

When we no longer look at an organic being as a savage looks at a ship, as at something wholly beyond his comprehension; when we regard every production of nature as one which has had a history; when we contemplate every complex structure and instinct as the summing up of many contrivances, each useful to the possessor, nearly in the same way as when we look at any great mechanical invention as the summing up of the labour, the experience, the reason, and even the blunders of numerous workmen; when we thus view each organic being, how far more interesting, I speak from experience, will the study of natural history become! (O 485–86)

The reference, again unspecified, is once again to Fuegians, about whom Darwin writes in the *Journal of Researches* that "[s]imple circumstances ... excited their admiration far more than any grand or complicated object, such as the ship. Bougainville has remarked concerning these very people that they treat the 'chef d'oeuvres de l'industrie humaine, comme ils traitent les loix de la nature et ses phénomènes' " [masterpieces of human industry as they treat the laws of nature and its phenomena] (J 189). If the citation of Bougainville suggests that Fuegians fail to comprehend the status of HMS *Beagle* as built rather than created, the product of human artifice rather than divine fiat, the "Conclusion" to the *Origin* explains this lack of comprehension as resulting specifically from an inability to see a ship as something "which has had a history," which is to say something with a meaningful and recoverable past. Conceiving of Fuegians as part of his past rather than possessed of their own, Darwin enlists what he imagines to be their blank indifference to the historicity of "grand or complicated" objects in the service of promulgating his conclusions about the past of all life. Further, Darwin attributes a similar indifference to his opponents, those benighted anti-evolutionists who mistakenly "look at an organic being as a savage looks at a ship."[44] In the context of the paragraph as a whole, the simile achieves a complicated pertinence. The reference to what might be called a savage way of looking enables a specific critique: so synonymous with amnesia and impercipience has the notion of the savage become that it can be employed figuratively to lambaste anyone who would deny the workings of natural selection or claim that evolutionary theory drains the biological world of mystery or interest. But the very terms of the critique demand, of course, that reasonable readers not be savages, that they refuse the savage in themselves. This delicate exchange, in which savages must be remembered in order for savage ways to be abandoned or forgotten, is repeated and intensified in that later and, if possible, still more controversial text, *The Descent of Man, and Selection in Relation to Sex.*

Having throughout the *Origin* left the question of human evolution to one side, Darwin in the *Descent* sets out directly to demonstrate "that man is descended from a hairy quadruped, furnished with a tail and pointed ears, probably arboreal in its habits" (*D* II: 389). In this text, that is, Darwin brings his great retrospective theory – which, in its refusal to view the natural world as a savage views a ship, invents or discovers a past for and a memory of life on the planet – to bear on himself and his readers. He is explicit about the mnemonic aspect of such an enterprise: "By considering the embryological structure of man, – the homologies which he presents with the lower animals, – the rudiments which he retains, – and the reversions to which he is liable," Darwin writes, "we can partly recall in imagination the former condition of our early progenitors; and can approximately place them in their proper position in the zoological series" (*D* II: 389).

Darwin's effort to "recall in imagination" his progenitors demands particular attention to the early stages of human existence, and, according to the dictates of the comparative method, those stages are illuminated with myriad references to so-called savages. A wide variety of remarks about their behavior and physical characteristics litters the text, as indicated by index entries ranging from "Bushwoman, extravagant ornamentation of a" (*D* II: 417) and "Malays, aversion of some, to hairs on the face" (*D* II: 445) to "Australians, colour of newborn children of" (*D* II: 409–10), "Fuegians, resistance of the, to their severe climate" (*D* II: 430), and "Savages, imitative faculties of … ; causes of low morality of …; uniformity of, exaggerated" and so on (*D* II: 461). What such entries do not quite indicate is the utter centrality of savages to Darwin's aims in the *Descent*, aims that he enumerates in the "Introduction" as follows: "to consider, firstly, whether man, like every other species, is descended from some pre-existing form; secondly, the manner of his development; and thirdly, the value of the differences between the so-called races of man" (*D* I: 2–3).[45] In their role as stand-ins for prehistoric forms of human life, savages provide key evidence for the solution to each of these three puzzles. They go to prove that humans are indeed descended from earlier forms of life; that they developed, like those earlier forms, according to the laws of natural and sexual selection; and that racial difference is the result, in Darwin's view, of the latter kind of selection.

Not until the very end of the final chapter of this mammoth text, however, in a retrospect titled "General Summary and Conclusion," does Darwin return again to the memory of his own initial experience of savages. The penultimate paragraph of the *Descent* begins as follows:

> The main conclusion arrived at in this work, namely that man is descended from some lowly-organised form, will, I regret to think, be highly distasteful to many persons. But there can hardly be a doubt that we are descended from barbarians. The astonishment which I felt upon first seeing a party of Fuegians on a wild and broken shore will never be forgotten by me, for the reflection at once rushed into my mind – such were our ancestors. These men were absolutely naked and bedaubed with paint, their long hair was tangled, their mouths frothed with excitement, and their expression was wild, startled, and distrustful. They possessed hardly any arts, and like wild animals lived on what they could catch; they had no government, and were merciless to every one not of their own small tribe. He who has seen a savage in his native land will not feel much shame, if forced to acknowledge that the blood of some more humble creature flows in his veins. (*D* II: 404)

At last the link between the memorability of savages and their status as representative of an earlier incarnation of humanity becomes explicit. The "for" in this passage carries causal force: the encounter with Fuegians remains permanently lodged in Darwin's memory because of

the realization that "such were our ancestors." At once impossibly different from Darwin and uncomfortably close to him, Fuegians constitute
a nightmarish double – as the term "reflection" suggests: a thought or
idea, "reflection" can also, of course, refer to the image produced by a
polished surface.[46] Both uses carry the root sense of turning back (Latin:
re-, back or again; *flectere*, to bend), as does the meaning, once possible
but most likely obsolete by the 1870s, "the recollection or remembrance
of a thing" (OED). Recoiling from his own image as if from a distorting mirror, Darwin reflects upon (returns to, considers, remembers) his
ancestors – and the reflection (thought, image, memory) horrifies. This
horror, though, can be put to use. In the rhetoric at play in this passage,
Darwin summons his reaction to Fuegians in order to ease acceptance of
the notion of human descent from other animals. Once Darwin's readers accede to the unpleasant but apparently self-evident fact that they are
descended from savages, that is, they should find it less difficult, indeed
in some sense a consolation, to admit that even further back in time their
ancestors were still more "lowly-organised" forms. As Darwin had written to Charles Kingsley nearly a decade earlier:

That is a grand & almost awful question on the genealogy of man to which you
allude. It is not so awful & difficult to me, as it seems to be most, partly from
familiarity and partly, I think, from having seen a good many Barbarians. I
declare the thought, when I first saw in T. del Fuego a naked painted, shivering
hideous savage, that my ancestors must have been somewhat similar beings, was
at that time as revolting to me, nay more revolting than my present belief that an
incomparably more remote ancestor was a hairy beast. Monkeys have downright
good hearts, at least sometimes, as I could show, if I had space.[47]

Still unforgettable, then, savages in the *Descent* return at first as that
which makes the kinship between humans and animals more palatable.
In the course of the penultimate paragraph, however, a subtle shift occurs
whereby Fuegians and Darwin's memory of his encounter with them move
from what enables acceptance of human descent from animals to what must
be refused and replaced by that descent. "For my own part," he writes,

I would as soon be descended from that heroic little monkey, who braved his
dreaded enemy in order to save the life of his keeper; or from that old baboon,
who, descending from the mountains, carried away in triumph his young com-
rade from a crowd of astonished dogs – as from a savage who delights to torture
his enemies, offers up bloody sacrifices, practises infanticide without remorse,
treats his wives like slaves, knows no decency, and is haunted by the grossest
superstitions. (*D* ii: 404–05)

Here Darwin asks readers to exercise their own faculties of memory in recalling an earlier moment in the *Descent*: the heroic monkey and loyal baboon alluded to are discussed in "Comparison of the Mental Powers of Man and the Lower Animals," the second and third chapters of the work's first volume. The burden of those chapters is to lessen the perceived distance between human and non-human with a barrage of anecdotes about animals' courage, affection, sociability, aesthetic sensibilities, capacity for reflection, and, not surprisingly, memory (*D* 1: 34–106).[48] At the end of the *Descent*, Darwin recurs to these anthropomorphized creatures in order to push "civilized" and "savage" humans further apart by grouping the former with animals against the latter, who now stand outside of any community whatsoever – in Peter Hulme's resonant formulation, "temporal castaways in the sea of modernity."[49] Belying the equivalence between animal and savage initially mooted ("I would as soon be descended from ...") and carrying an affective charge far in excess of the requirements of the analogical argument being prosecuted, the mention of savages and the demonization of savage behavior memorialize a deep antipathy to the thought of a kinship of which, nevertheless, "there can hardly be a doubt."

4

Half a century after Darwin's encounter with Fuegians, when W. H. Hudson writes about his own travels in the southern cone of South America, he does so with Darwin constantly in mind. It might even be said that Darwin is to Hudson as Fuegians are to Darwin insofar as Hudson cannot help but recall Darwin at crucial moments but also repeatedly attempts to refuse connection with him. Such refusal begins on his title page itself, which announces the intention to document neither the heroic, active, Odyssean *Voyage of the Beagle* nor the empirical encounters that form the basis for a sober *Journal of Researches* but instead *Idle Days in Patagonia*. Hudson is more like than unlike the author of the narrative of the *Beagle*'s voyage, though, insofar as a concern with memory suffuses his text as well, coloring the whole – or scenting it, it might be better to say, since the olfactory preoccupies Hudson much more than does the visual. The final chapter of *Idle Days in Patagonia*, "The Perfume of an Evening Primrose," conducts a nearly Proustian investigation of that titular perfume, its evocation of various pivotal events in Hudson's life, and the more general claim that among the senses it is smell that can best "restore the past" because, unlike sights and sounds, scents "cannot ... be reproduced in the mind" – cannot, that is, deliberately be remembered.⁵⁰ Despite the emphasis here on the particular ability of scents to effect such time traveling, a similar restoration of days gone by occurs elsewhere in the text as a result of landscape. As we have seen, the plains of Patagonia function in much the same way as the perfume of the evening primrose, providing a mnemonic device so powerful that it obliterates the present, replacing it not with the personal past of a lost childhood but with the collective past of "the emotional state of the pure savage."

In connection with Darwin and Fuegians, the key point about Hudson's reversion to savagery is neither its mere occurrence nor the particular stimulus that instigates it but rather the response to which it gives rise: "the feeling experienced on going back to a mental condition we have outgrown, which I had in the Patagonian solitude," Hudson announces, is nothing other than "elation."⁵¹ Traveling backwards in time so as to retrace and reverse the path of human evolution, Hudson arrives at what he takes to be the stark clarity and alertness of the savage mind with excitement, welcome, relief – the emotions of homecoming. The response is hardly unprecedented. As Torgovnick notes, "[t]he metaphor of finding a home or being at home recurs over and over [in] Western primitivism. Going primitive is trying to 'go home' to a place that feels comfortable and balanced, where full acceptance comes freely and easily."⁵² But such yearning for a lost sense of belonging both structures and is rendered curiously obsolete by *Idle Days in Patagonia*, a text in which the chasm separating past from present, primitive from civilized, is at once insisted on and effortlessly bridged over. If such a chasm amounts to thousands or tens of thousands of years, that apparently insuperable temporal distance becomes inconsequential on the plains of Patagonia, disappearing to reveal the joyful truth that "in our inmost natures, our deepest feelings, we are still one with the savage."⁵³

Hudson provides an especially intense but nonetheless exemplary instance of the romanticization of what are imagined to be savage ways of life and states of being – the welcoming embrace of the savage that stands counter to Darwin's fear and loathing. At the same time, despite his differences from Darwin, he helps elucidate the double and contradictory place of the savage in Darwin's work. For if, as Darwin writes at the end of the *Descent*, "there can hardly be a doubt that we are descended from barbarians," the likelihood that, as Hudson puts it in *Idle Days in Patagonia*, "we are still one with the savage" renders such indubitable knowledge insupportable. Incapable of exorcising the memory of his own encounter with Fuegian "savages" and, further, compelled by the exigencies of his argument to invoke that encounter to demonstrate the workings of evolution, and human evolution in particular, Darwin denies the savage in order to render the consequences of those workings tolerable. Certain of his kinship with savages, he nevertheless betrays a wish that he did not have to recall it. Hudson appears to reveal the motivation behind that wish as fear of atavism, of a literal return to savagery, and Darwin's corpus amply attests to such a fear. If we look back at the *Autobiography*, for instance, we find a passage explaining how, during the first two years of the *Beagle*'s voyage, Darwin gradually turned over responsibility for shooting specimens to his servant so that he might spend all his time tagging, cataloguing, and geologizing, a passage that Darwin concludes with the observation about himself that "[t]he primeval instincts of the barbarian slowly yielded to the acquired tastes of the civilized man."[54] Separated by only a paragraph from the account of the unforgettables of the *Beagle* voyage, and so from the mention of "[t]he sight of a naked savage in his native land," the confident progressivism of this sentence remains vulnerable to reversal, to the slow yielding (or, as in Hudson, instantaneous reversion) of acquired tastes to primeval instincts, of the civilized man to the barbarian.

Darwin might be said actually to have undergone such an atavistic reversal, and to have done so by way of the very forgetfulness he summons up to ward off that possibility. For among the various elements that constitute humanity for Darwin is, above all, the capacity to reflect on past actions and events, a capacity wholly dependent on the faculty of memory. Only in and through their employment of memory and reflection do humans as such, as ethical beings able to evaluate and alter their own behavior, exist. "A moral being," he writes in the *Descent*, "is one who is capable of comparing his past and future actions or motives, and of approving or disapproving of them" (*D* 1: 88). And a page further on: "Man, from the activity of his mental faculties, cannot avoid reflection: past impressions and images are incessantly passing through his mind with distinctness" (*D* 1: 89). Hence the significance of Darwin's repeated claims that non-human animals, too, have the ability to remember and reflect: if we are descended from them, they must possess, if only partially or incipiently, even those "higher" capacities (in this case, an ethical sense and the powers of recollection on which it depends) that appear to set us apart from them.[55]

And so, again in the chapters of the *Descent* in which he undertakes a "Comparison of the Mental Powers of Man and the Lower Animals," Darwin poses the following patently rhetorical question:

But can we feel sure that an old dog with an excellent memory and some power of imagination, as shewn by his dreams, never reflects on his past pleasures in the chase? and this would be a form of self-consciousness. On the other hand ... how little can the hard-worked wife of a degraded Australian savage, who uses hardly any abstract words and cannot count above four, exert her self-consciousness, or reflect on the nature of her own existence. (*D* I: 62)

What distinguishes savages from the civilized, and apparently even from "an old dog with an excellent memory and some power of imagination," is their inability to reflect, itself rooted in an inability to remember.[56] Fuegians display such an inability, as, in a different register, do the savage Britons who would oppose the theory of evolution. In incessantly recalling Fuegians and his encounter with them, then, Darwin enacts his own civilized humanity. As for so many other Victorian narrators, his engagement in retrospection makes him who he is. Looking back, he pieces together a story of origins and development that, because of its totalizing ambitions and the scope of its influence, becomes for a time *the* story of development, the privileged account by which nineteenth-century Europeans understood themselves and how they came to be. Troubling that account, though, is the danger of a constitutive forgetting that, even as it works to distance past from present, savage from civilized, threatens to instill savagery in the modern. In wishing to ward off as much as to recall his proximity to Fuegians, that is, Darwin leaves open the possibility of succumbing to the amnesia and incomprehension he condemns. Paradoxically desiring to look at savages in the way he claims savages look at ships and anti-evolutionists look at the natural world, Darwin risks becoming one of the most "civilized" beings conceivable, the modern primitive.

Despite, however, the gestures of avoidance that run from the *Journal of Researches* ("could our progenitors have been men such as these?" [*J* 474]) to the *Descent* ("I would as soon be descended from that heroic little monkey" [*D* II: 404]), Darwin never comes close to forgetting Fuegians. For how could he? His entire project demands that he remember, and that we remember with him, what he may not wish to and in a literal sense, *pace* Hudson, cannot: the history of life as an unfathomably long series of imperceptibly minute changes in millions of generations of millions of species of biological beings, including those human ancestors for whom, for Darwin and other practitioners of the comparative method, nineteenth-century savages stand. Thus, to return to and reinflect a notion advanced earlier in this chapter, Darwin's corpus may not resemble "one long argument" so much as one immense prosthetic memory. In the *Principles of Geology*, Lyell likens the geological record to a book in which, despite various missing pages, the earth's past can be discerned.[57] Darwin himself borrows and extends the analogy. "For my part," he writes,

following out Lyell's metaphor, I look at the natural geological record, as a history of the world imperfectly kept, and written in a changing dialect; of this history we possess the last volume alone, relating only to two or three countries. Of this volume, only here and there a short chapter has been preserved; and of each page, only here and there a few lines. (*O* 310–11)

But Darwin's textual production taken as a whole requires that the metaphor be reversed: geological strata reconstructed in and as language, like those strata it remembers for us what we are incapable of remembering ourselves.

The imperative to remember characterizes Darwin's work in another way as well, one that will return us for the last time to Darwin's persistent and persistently resisted memories of Fuegians. The subtitle of the *Origin* defines evolution by natural selection as "The Preservation of Favoured Races in the Struggle for Life," and it is this view of nature as an arena of perpetual conflict that, for better and worse, has come to be indissolubly associated with Darwinism. For Darwin himself, however, accepting evolution as a valid account of biological change always implied a different kind of struggle, the struggle against forgetting. In a passage in the *Origin* that attempts nothing less than a thoroughgoing redescription of the natural world, Darwin writes:

We behold the face of nature bright with gladness, we often see superabundance of food; we do not see, or we forget, that the birds which are idly singing round us mostly live on insects or seeds, and are thus constantly destroying life; or we forget how largely these songsters, or their eggs, or their nestlings, are destroyed by birds and beasts of prey; we do not always bear in mind, that though food may be now superabundant, it is not so at all seasons of each recurring year. (*O* 62)

"[W]e forget … we forget … we do not always bear in mind": without the necessary corrective of memory, immediate impressions of nature "bright with gladness" inevitably deceive. Beer beautifully limns this aspect of Darwin when she writes that he "lives in a doubly profuse world – the plenitude of present life, its potential for both development and death, and the recessional and forgotten multitudes which form the ground of the present."[58] It is on the possibility of recalling those forgotten multitudes, he insists, that our ability to comprehend evolution depends.

Thus the justly celebrated "entangled bank" paragraph that concludes the *Origin*, so often invoked as emblematic of Darwinian nature, provides at best a partial account. What is seen in that paragraph, "elaborately constructed forms" that include "plants of many kinds," "birds singing on the bushes," "various insects flitting about," and "worms crawling through the dark earth," can be fathomed only in relation to what is not seen, to what is absent and therefore – here is where what might seem a truism takes a surprising turn – must be remembered (*O* 489). The *Journal of Researches*, the *Origin*, the *Descent*, the *Autobiography*: each insists that there can be no understanding without recollection, no discovery without memory. More particularly, each aspires to serve as a reminder (as the passage above serves as a reminder) of that which is easiest to forget but also, if evolution by means of natural selection is to be grasped, essential to recall: the ubiquity of conflict, loss, death.

As Adam Phillips puts it in *Darwin's Worms*: "One is, Darwin suggests … more likely to forget what is disagreeable to oneself, but what is most disagreeable may be of most value."[59] But this formulation, precisely accurate with regard to the place of starvation, predation, and extinction in Darwin's corpus, nonetheless strangely fails when it comes to Fuegians. Undoubtedly "disagreeable" and arguably of "most value" in thinking about humans and their origins, Fuegians are not simply unlikely to be forgotten by Darwin but actually impossible for him to forget. They thus constitute an exception to or short-circuit in the work of memory demanded by evolutionary theory. To explain why this should be so, let me invoke once more a claim made near the end of the third section of this chapter: namely, that Darwin anthropomorphizes non-human animals in order to group them with civilized humanity against savages, who are thus effectively exiled, left without a place in the web of life. (Recall the *Journal of Researches*: "Viewing such men, one can hardly make oneself believe they are fellow-creatures, and inhabitants of the same world" [*J* 184].) Such exile, I have argued, constitutes a refusal of the proximity of the savage to the civilized that is at once necessary and unbearable. But there is another, equally compelling, and, I think, more suggestive way to conceptualize the displacement of the savage, one that has to do with the fate of the "human" under evolution.

In the wake of Darwinism, the distinction humans make between themselves and other animals must be seen as essentially arbitrary. Like "species" in the *Origin* ("Certainly no clear line of demarcation has as yet been drawn between species and sub-species … or, again, between sub-species and well-marked varieties" [*O* 51]), "human" after the *Descent* becomes a term of convenience, a name for a difference that may not actually exist.[60] As evidentiary and rhetorical constructs, Fuegians function to render this insight possible. Darwin's repeated references to and incessant memories of them as inhuman or as occupying the borderland between human and inhuman enable the postulation of a continuum running from the most "lowly-organized form[s]" of animal existence to civilized humanity – a continuum that can be put in motion and read as a time-line that tells the story of life. Necessarily existing at a point on that continuum, Fuegians are nonetheless also exiled from it. Neither human nor "animal," too familiar to be ignored but too alien to be acknowledged kin (and, as Ian Duncan argues, vice versa), they stand alone.[61] Because of this very isolation, however, Fuegians remain as a kind of last possibility for human exceptionalism in Darwin's evolutionary theory. At a moment when other humans have been completely assimilated to the natural world, when the line between human and nonhuman has been revealed to be a consolatory fiction, Fuegians preserve human separateness. Whence their place, and the place of savages more generally, in the play of memory and forgetting requisite to theorizing evolution: if understanding natural selection requires a struggle against forgetting what is disagreeable, the involuntary memory of Fuegians serves to forestall one of the most disagreeable consequences of evolution, the loss of human distinctness. As attested by the joyful embrace of the savage within evinced by Hudson and so many others since, Darwin's memory of "savages" on a wild and desolate shore should be understood as, among other things, a memory of the last fully human beings on earth.

CHAPTER 2

Alfred Russel Wallace's tropical memorabilia

Alfred Russel Wallace's *A Narrative of Travels on the Amazon and Rio Negro* (1853) is a work born of loss and, consequently, a work of memory. On August 6, 1852, the *Helen*, the ship carrying Wallace home to England after four years in South America, caught fire and sank. With it went half the journals he had kept, together with most of his sketches and collections.[1] Wallace's description of the ten days subsequently weathered in a small open boat 700 miles from the nearest land is brief and matter of fact, its single strongly marked emotion reserved for the moment of rescue: "We were saved!" (*N* 400). Only in retrospect, safe on board the cargo ship *Jordeson*, does he think of what he has suffered.[2] Looking back on his ordeal, he does not recall the two years' worth of notes that went down in the Atlantic, the pain of thirst and exposure, or even the fear of dying at sea. He recalls his irrecoverable specimens:

> It was now, when the danger appeared past, that I began to feel fully the greatness of my loss. With what pleasure had I looked upon every rare and curious insect I had added to my collection! How many times, when almost overcome by the ague, had I crawled into the forest and been rewarded by some unknown and beautiful species! How many places, which no European foot but my own had trodden, would have been recalled to my memory by the rare birds and insects they had furnished to my collection! How many weary days and weeks had I passed, upheld only by the fond hope of bringing home many new and beautiful forms from those wild regions; every one of which would be endeared to me by the recollections they would call up, – which should prove that I had not wasted the advantages I had enjoyed, and would give me occupation and amusement for many years to come! And now everything was gone, and I had not one specimen to illustrate the unknown lands I had trod, or to call back the recollection of the wild scenes I had beheld! (*N* 400–01)

Introduced as a substantiation of "the greatness of [Wallace's] loss," the passage quickly modulates into a melancholy reflection on the joys of collecting. The frequent exclamation marks – five of these six sentences end with one – well convey the tonal and temporal complexity of the whole, for they signify at once the initial excitement of discovery and the subsequent realization of the loss of the things discovered. "With what pleasure," Wallace writes, "had I looked upon every rare and curious insect I had added to my collection!" – exclaiming at the pleasure as well as at having been robbed of its embodiment. The anaphora that follows itemizes the occasions of collecting by frequency ("How many times"), location ("How many places"), and duration ("How many weary days and weeks"), establishing the intensity of joy and loss and, at the same time, specifying the work done by the natural-historical collection as the binding together of the temporal and the spatial. The items collected link past times and exotic places to one another. Moreover, because of their portability, they provide the means to link those times and places, in turn, to the future and to England: the "fond hope" that sustains Wallace through illness and exhaustion is one of "bringing home many new and beautiful forms from those wild regions." But such portability also means vulnerability. Thus it happens that, after the sinking of the *Helen*, "everything was gone": not specimens alone but, with them, past pleasure and future reward; proof and hope; rarity, novelty, and beauty.

Memories, too, have gone missing. Curiously, Wallace recalls having been deprived of opportunities to remember. Promising "to illustrate the unknown lands [he] had trod, to call back the recollection of the wild scenes [he] had beheld," his specimens were to have been exemplary and mnemotechnic. Lacking them, he lacks access to his own past – a past that was to have provided "occupation and amusement for many years to come." He recollects what he has lost as that which would have called forth recollection, mourning absent things because of their relation to the absent places and feelings they were to have commemorated. Thus Wallace's memory of loss turns out to be a meditation on the loss of memory.[3] As so often in his work, what had seemed assured, solid – pinned insects, skinned and stuffed birds, a small menagerie of "parrots, monkeys, and other [live] animals we had on board" – turns out to have been fragile (*N* 395). But more fragile still is the memory that might have been capable of effecting repair. Called upon to restore the things destroyed, that memory can only summon up their loss as the moment it, too, suffered destruction. In this way Wallace reveals one of the chief risks of the inter-implication of collecting and recollecting: if natural-historical anamnesis requires a material prop, the cost of the disappearance of memorabilia is forgetting.

Having accounted for the genesis of *A Narrative of Travels* in losses remembered and memories lost, Wallace concludes the book with a memory that survives and a prediction of future loss made bitter by that memory. The third paragraph of the book's final chapter, "On the Aborigines of the Amazon," begins with a definitive retrospective judgement:

I do not remember a single circumstance in my travels so striking and new, or that so well fulfilled all previous expectations, as my first view of the real uncivilized inhabitants of the river Uaupés … I felt that I was as much in the midst of something new and startling, as if I had been instantaneously transported to a distant and unknown country. (*N* 477)

Recollections of wild scenes and places untrodden by other European feet, bound up with particular specimens, disappear with those specimens' disappearance. But the memory of the "first view of the real uncivilized inhabitants" of South America remains. Perhaps that memory is self-sustaining because what Wallace elsewhere calls "savages" constitute for him the most novel aspect of his journey. Perhaps, on the other hand, savages could not be forgotten for what might appear to be the opposite reason: because they "so well fulfilled all previous expectations." I will have occasion later in this chapter to address the paradox of how something absolutely new can be, at the same time, absolutely expected. The point at the moment is simply that Wallace's memory of the savage requires no material armature. Unlike those places and times commemorated by collection that, as a result, prove impossible to recall without recourse to specimens, the first view of indigenous South Americans cannot be forgotten. For Wallace no less than for Darwin, savages memorialize themselves.

The rest of "On the Aborigines of the Amazon" moves away from questions of the sheer memorability of the peoples of the Amazon basin to provide a digest of ethnographic observations about their physical appearance, geographical distribution, and characteristic tools, dwellings, and foods. Turning at chapter's end to speculation on the future of these peoples, Wallace envisions the possibility that such "ingenious and skilful workmen," who "have, and show, a great affection for their children" and "scarcely ever quarrel among themselves," may be "formed, by education and good government, into a peaceable and civilized community" (*N* 518–19). But the final paragraph of the chapter – which, excepting the appendix, is also the final paragraph of the book as a whole – offers up a darker prospect: "This change however will perhaps never take place: they are exposed to the influence of the refuse of Brazilian society, and will probably, before many years, be reduced to the condition of the other half-civilized Indians of the country, who seem to have lost the good qualities of savage life, and gained only the vices of civilization" (*N* 519). Remembering savages with admiration and attempting to think of their future as a bright one, Wallace concludes with a prediction of their imminent destruction at the hands of a civilization that ruins what he finds admirable about their life and imparts in return only new vices.[4] Savages, unlike wild scenes and untrodden places, may be remembered without the aid of a material mnemonic, but such memory brings with it little hope or amusement. It constitutes a fond glance back that includes, as an apparently inevitable corollary, a harrowing look forward to the probable disappearance of what it recalls.

Specimens and savages, memory and loss, retrospection and prognostication: the insistent presence of these concerns characterizes not *A Narrative of Travels* alone but the whole of Wallace's life and work. For, to begin with, the search for specimens and the study of savages were connected. Wallace undertook both with a specific aim in mind. As he wrote to Henry Walter Bates just before the two departed for the Amazon together, he wished to gather data in the tropics "principally with a view to the theory of the origin of species."[5] Sympathetic to the transmutationist thesis put forward in Robert Chambers's anonymously published *Vestiges of the Natural History of Creation* (1844), Wallace set out for South America in late April, 1848, actively seeking evidence to confirm that thesis.[6] His subsequent voyage from 1854 until 1862 throughout the chain of Southeast Asian islands then known as the Malay Archipelago was a continuation of the same search. Wallace traveled, collected, and made natural-historical as well as ethnographic observations with the express purpose of solving the riddle of how species come into being, a riddle that he did solve on the island of Gilolo in February, 1858. The paper he wrote that month and mailed to Darwin, "On the Tendency of Varieties to depart indefinitely from the Original Type," famously spurred the reluctant older man to action: by November of the following year, after more than two decades of keeping the theory of evolution by natural selection to himself and a few trusted confidants, Darwin had composed an abstract of the *magnum opus* he was in the process of writing, *Natural Selection*, and published it under the title *On the Origin of Species, or, The Preservation of Favoured Races in the Struggle for Life* (1859).[7]

Just as, in the *Origin*, Darwin refrains from directly asserting that humans had come into existence in the same way that other living beings had, Wallace makes no reference to human evolution in "On the Tendency of Varieties to depart indefinitely from the Original Type." But the "view to the theory of the origin of species" that Wallace mentioned to Bates was a view that had always taken in *Homo sapiens*. The peoples he encountered and studied in the Amazon and, later, in Southeast Asia proved indispensable to his thinking as he worked out how species replace one another through time according to what he refers to in the 1858 paper as the law of *"progression and continued divergence."*[8] Moreover, as Martin Fichman observes: "Whatever Wallace's fascination with beetles and butterflies, the human implications of evolution were always in the foreground of his thought."[9] In the years following the publication of his and Darwin's theory, it was Wallace who formulated one of the earliest significant treatments of the consequences of that theory for understanding humans, "The Origin of Human Races and the Antiquity of Man Deduced from the Theory of Natural Selection" (1864).[10]

As is well known, the irony is that it was Wallace, too, who not long afterwards abandoned the belief that humans could have been produced by natural selection alone. But it is not frequently noted that indigenous peoples provided him with what he took to be the decisive evidence against that belief. In an 1869 review of the tenth edition of Charles Lyell's *Principles of Geology* (1868), Wallace avers that the capacity of savages' brains and the perfection of their bodily form exceed what is required for their survival and so run counter to the Darwinian dictum that natural selection can only give rise to structures immediately useful to their possessors.[11] If natural selection could not have produced the savage brain and body unaided, it must have had supernatural help. Considered in this light, he reasons, savages prove the existence of "an Overruling Intelligence" that has guided the process of human evolution, "so directing variations and so determining their accumulation, as finally to produce an organization sufficiently perfect to admit of, and even to aid in, the indefinite advancement of our mental and moral nature" ("Lyell," 394). A decade after he had mailed his fateful paper "On the Tendency of Varieties to depart indefinitely from the Original Type" to Darwin, the independent co-founder of the theory of evolution by means of natural selection reversed his position on humans – because, in large part, of what he remembered about savages.

For Wallace, then, as for Darwin, savages constitute an exception. Darwin, in a deft negotiation between involuntarily remembering and strategically forgetting the savage, sutures over that exception in order to retain an unbroken evolutionary continuum. Wallace, by contrast, embraces it and the rupture it entails. Encounters with savages initially provide some of the data that allowed him to formulate the theory of evolution. But his memories of those encounters soon come to mark the point at which, for him, that theory breaks down. Those memories thus prove essential to his rejection of unaided evolution as sufficient to explain the appearance of humans on the earth.[12] In addition to serving as one of the causes of his evolutionary apostasy, they also provide him with a point of vantage from which to launch a thoroughgoing critique of "civilization." Darwin, viewing Fuegians as the incarnation of forgetfulness and incomprehension, attacks the opponents of evolutionary theory by troping them as savages. Wallace, finding in some of the inhabitants of Amazonia and the Malay Archipelago examples of physical perfection living in social harmony, levels the charge of savagery at Victorian Britain in its entirety. The contrast between savage and civilized life, according to Wallace, reveals the true barbarism of the latter.[13] Finally, Wallace's savage mnemonics evidences an interplay between memory and evolutionary theory significantly different from that found in Darwin's work. Like the memories Darwin insists must constantly be borne in mind if we are to understand natural selection, Wallace's memories of savages often have to do with loss: loss of certainty, beauty, possibility. Anticipating the exoticist nostalgia of Claude Lévi-Strauss's *Tristes tropiques* and so many similar texts, Wallace recalls the savage as something valuable in the process of disappearing.[14] But the recollection the savage affords Wallace is not of a lost world from which the modern is forever shut out, an elsewhere to which one might have escaped, once upon a time. Rather, it is of the ethical and social heights to which humans once evolved – and to which they might again aspire. Insisting on an ultimate teleology for evolutionary change, Wallace at the same time, via his memory of savages, upholds the radically nonteleological implications of Darwinian theory: that movement forward in time does not necessarily imply progress.[15]

I

It would be difficult to find two autobiographies that differ from one another more than Wallace's *My Life* (1905) differs from Darwin's *Autobiography* (1887). The two thick volumes of *My Life* cover an immense amount of ground: Wallace's childhood and youth, travels, friendships, socialist politics, spiritualist beliefs, and contributions to science. Rich in details, the book includes excerpts from Wallace's correspondence and summaries of important publications. Such dilation is worlds away from Darwin's spare and piecemeal "Recollections of the Development of [his] mind and character."[16] At one point, however, Wallace seems to have taken his script directly from Darwin. For just as Darwin, reflecting on his travels on the *Beagle*, singles out the three aspects of the voyage that remain clearest in his memory as he looks back from near the end of a long life, so Wallace sums up the years he spent in South America half a century before by listing "the three great features which especially impressed [him], and which fully equalled or even surpassed [his] expectations" (*ML* I: 286–87). Repeating Darwin's wonder at the "glories of the vegetation of the Tropics," Wallace lists "the virgin forest" as the first of those features. Next, in the position Darwin awards to the deserts of Patagonia, Wallace places the "wonderful variety and exquisite beauty of the butterflies and birds" (*ML* I: 287). Finally, in his closest echo of the *Autobiography*, Wallace reserves for the last and most remarkable aspect of his journey his encounter with savages. Darwin had written: "The sight of a naked savage in his native land is an event which can never be forgotten."[17] Wallace, characteristically more prolix, writes:

The third and most unexpected sensation of surprise and delight was my first meeting and living with man in a state of nature – with absolute uncontaminated savages! This was on the Uuapés [*sic*] river, and the surprise of it was that I did not in the least expect to be so surprised. I had already been two years in the country, always among Indians of many tribes; but these were all what are called tame Indians, they wore at least trousers and shirt; they had been (nominally) converted to Christianity, and were under the government of the nearest authorities; and all of them spoke either Portuguese or the common language, called "Lingoa-Geral."

But these true wild Indians of the Uaupés were at once seen to be something totally different. They had nothing that we call clothes; they had peculiar ornaments, tribal marks, etc.; they all carried weapons or tools of their own manufacture; there were living in a large house, many families together, quite unlike the hut of the tame Indians; but, more than all, their whole aspect and manner were different – they were all going about their work or pleasure which had nothing to do with white men or their ways; they walked with the free step of the independent forest-dweller, and, except the few that were known to my companion, paid no attention whatever to us, mere strangers of an alien race. In every detail they were original and self-sustaining as are the wild animals of the forests, absolutely independent of civilization, and who could and did live their own lives in their own way, as they had done for countless generations before America was discovered. I could not have believed that there would be so much difference in the aspect of the same people in their native state and when living under European supervision. The true denizen of the Amazonian forests, like the forest itself, is unique and not to be forgotten. (*ML* I: 288)

Despite its greater length, Wallace's account shares much with Darwin's. Darwin, employing alliterative emphasis, describes his initial sight of a Fuegian as an encounter with something unadulterated: a "naked savage in his native land," readers are given to understand, embodies the *ne plus ultra* of savagery. Wallace, too, singles out what he takes to be the purity of the inhabitants of the banks of the Uaupés, marking its import with an exclamation point: they are for him "absolute uncontaminated savages!" Nakedness, apparently in part as a sign of such lack of contamination, also figures in both accounts – even if Wallace's relativizing comment that "[t]hey had nothing that we call clothes," a comment further qualified by the immediate mention of "peculiar ornaments, tribal marks, etc.," departs from Darwin's starkly rendered "naked savage."[18] Finally, Wallace, like Darwin, describes South American indigenes as, above all, unforgettable. For Darwin it is specifically the "sight" of a savage that constitutes an "event which can never be forgotten"; for Wallace savages themselves, "true denizen[s] of the Amazonian forests," are "unique and not to be forgotten."[19] Nonetheless, both signal the particular nature of their encounter by way of its indelibility, its fixity in their memory.

But precisely here, with memory, the parallels between these two retrospective narratives come to an end. For Wallace's reasons for remembering savages are different from – in fact, nearly the opposite of – Darwin's. Establishing the full extent of this difference requires the invocation of another Darwinian intertext, the *Journal of Researches into the Natural History and Geology of the Various Countries Visited by H. M. S. Beagle* – which Wallace refers to in *My Life* as one of the two books that instilled in him a desire to visit and study the tropics (*ML* 1: 256).[20] In the final chapter of the *Journal of Researches*, Darwin takes stock of the remarkable things he saw on his voyage, noting: "Of individual objects, perhaps no one is more certain to create astonishment than the first sight in his native haunt of a real barbarian, of man in his lowest and most savage state."[21] Repeating Darwin's declaration of "astonishment," Wallace calls his own experience a "most unexpected sensation of surprise." But that surprise is not, as it had been for Darwin, accompanied by dismay. Instead, Wallace feels a pleasure so intense that he calls it "delight."

The rest of the passage in *My Life* indicates the causes of that delight, which have to do primarily with what Wallace sees as the autonomy of the people he encounters. They impress him because of their clearly demonstrated sense of separateness, conveyed, in part, by their indifference to him and his companion, "mere strangers of an alien race" who have appeared among them.[22] "[O]riginal" and "self-sustaining," they "could and did live their own lives in their own way." Such self-possession is what makes them "unique," and Wallace's delight in it what makes them "not to be forgotten." The indelibility of Darwin's memory of the savage derives from shock closely related to horror; that of Wallace's, from surprise given force by pleasure. Rescripting Darwin's account of the kind of affective response that leads to memorability, Wallace rescripts, too, his own meditation on the workings of memory in the passage about the sinking of the *Helen* with which I began. There the pleasures of collecting specimens cannot provide a guarantee against the amnesia that accompanies their loss; here the delight attendant on an encounter with savagery ensures that the savage cannot be forgotten.[23] Such delight may seem, and in one sense surely is, the xenophilic complement of Darwin's xenophobic recoil from Fuegians – an apparently distinct but in fact closely related form of exoticism. As I hope to show, however, this delight makes a difference. Indeed, it may be said to make all the difference insofar as it proceeds from and in turn enforces a sense of fundamental commensurability between "civilized" and "savage" that forms the basis for Wallace's subsequent theorizing, including his contentions about the barbarism of what passes for civilization, the unity of the human species, and the possibility of evolutionary decline as well as advance.

A fuller sense of the contours of such commensurability is to be found in the section of *A Narrative of Travels on the Amazon and Rio Negro* in which Wallace first publicly narrates his initial experience of "unadulterated" savages. On June 7, 1851, he relates, while traveling up the Río Uaupés, a tributary of the Río Negro, he and his small party stop at "a 'malocca,' or native Indian lodge, the first we had encountered" (*N* 275). He begins his account of what he saw by calling attention to the same feeling he would later emphasize in *My Life*: delight. "On entering this house," he declares, "I was delighted to find myself at length in the presence of the true denizens of the forest" (*N* 277). Observing that the women, usually naked, pull on clothing when he enters the longhouse, he goes on to state: "It was the men however who presented the most novel appearance, as different from all the half-civilized races among whom I had been so long living, as they could be if I had been suddenly transported to another quarter of the globe" (*N* 277).

Even at this early moment in Wallace's depiction of South American indigenes it is clear that his response to them has something distinct about it. Most notably, he does not represent the people in whose presence he finds himself as if they inhabited an earlier time than he does. Quite different from Darwin and many other nineteenth-century European travelers to South America (and elsewhere as well), Wallace writes as though he and these "true denizens of the forest" share the same historical moment. He does not figure them as his ancestors or as providing any clue to what such ancestors may have been like. What sets them apart comes into view when he compares them, not to himself, but rather to what he calls the "half-civilized" Brazilians and Venezuelans among whom he had been traveling for nearly three years.[24] Eschewing the denial of coevalness essential to the workings of the comparative method, Wallace turns to a geographical rather than a temporal figure of speech: he feels, he says, as if "transported to another quarter of the globe."[25] Darwin remembers savages as themselves memories and reminders, visitors from his own pre-history. In this passage, Wallace depicts them as giving rise to a sense of dislocation in space rather than time. No history, theirs or his, enters in; instead, the intensity of the present moment is conveyed by way of a simile likening its effect to a feeling of vertiginous displacement. Although in many subsequent discussions Wallace will treat indigenous peoples as exemplary of prehistoric humans, the force of this initial sensation of spatial rather than temporal shock is definitive insofar as it returns throughout his work to trouble any sense of his and other Europeans' superiority, and particularly any notion of savages as outmoded or unevolved.

One reason for the differences between Darwin's and Wallace's accounts may have to do with the corresponding differences in the peoples they encounter. Each writes of viewing "savages," but the single term misrepresents insofar as it names distinct groups of humans living in distinct environments. (Of course, its misrepresentation is not limited to this false generality.) The lush forests, warm climate, and plentiful food of Amazonia could hardly be further from the frigid and barren shores of Tierra del Fuego. Equally removed from each other are the social arrangements of the peoples living along the Uaupés as described by Wallace and those of Fuegians as reported by Darwin. Such differences undoubtedly account in some measure for these two men's disparate responses.

But the distinctness of those responses also derives, in at least equal measure and with more consequence, from differences of approach. Consider, in this regard, Darwin's reference to the "first sight in his native haunt of a real barbarian," which insists on the distance between "civilized" and "savage" by figuring their relation as purely visual, one between observer and observed. Wallace collapses that distance. In *My Life*, he writes: "The third and most unexpected sensation of surprise and delight was my first meeting and living with man in a state of nature – with absolute uncontaminated savages!" It is "first meeting and living with" such people, not simply a "first sight" of them, that gives rise to the sensation he remembers and records. Thus savages are not, for Wallace, to be confused with specimens. One can glimpse a first sight of a bird or a butterfly; one cannot, at least in the usually accepted sense of the terms, meet one, still less live with one.[26] For Darwin, as I demonstrate in the previous chapter, if Fuegians are in an important sense neither animal nor human, in another sense they stand as the only authentic humans remaining in a post-evolutionary world. They occupy a liminal space in his thought that allows alternately for their demonization and valorization. Either way, only in the Anglicized and Christianized form of Jemmy Button and his fellow kidnap victims can they be met. For Wallace, on the other hand, while savages are arguably animal – existing, as he writes, "in a state of nature" – they are also indisputably human – since, even in such a state, one can meet them and, more remarkably, live with them.[27]

To put this another way, one crucial source of Wallace's delight in and memory of the ways of savages is also an aspect of what makes them like himself: the possibility of intimacy with them. So, to return to *A Narrative of Travels*, the sense of spatial dislocation to which savages give rise is literalized (but also assuaged) by Wallace's relocation to an unfamiliar domestic interior. The continuation of the depiction of the initial moment of encounter figures such intimacy by way of the additional detail that Wallace and his companions remained in the longhouse until the next morning. Of this stay he notes, "We passed the night in the malocca, surrounded by the naked Indians hanging round their fires, which sent a fitful light up into the dark smoke-filled roof. A torrent of rain poured without, and I could not help admiring the degree of sociality and comfort in numerous families thus living together in patriarchal harmony" (*N* 279–80).[28] To live with savages, Wallace gives us to understand, is to be privy to revelations about their way of life otherwise inaccessible, revelations that then prove indelible. Looking back on this moment years later, he will be unable to forget what he regards as the idyllic social organization that characterizes the communal living beneath the roof of the malocca.

The settlement Wallace visited on June 7 was relatively empty of inhabitants; most, he reports, had left for a feast being held nearby. Two days later, Wallace arrived at the village hosting the feast, an event that provides the occasion for another, complementary but quite different rendering of memorable savagery:

On entering the great malocca a most extraordinary and novel scene presented itself. Some two hundred men, women, and children were scattered about the house, lying in the maqueiras [hammocks], squatting on the ground, or sitting on the small painted stools, which are made only by the inhabitants of this river. Almost all were naked and painted, and wearing their various feathers and other ornaments. Some were walking or conversing, and others were dancing, or playing small fifes and whistles … The wild and strange appearance of these handsome, naked, painted Indians, with their curious ornaments and weapons, the stamp and song and rattle which accompanies the dance, the hum of conversation in a strange language, the music of fifes and flutes and other instruments of reed, bone, and turtles' shells, the large calabashes of caxirí [beer made from manioc] constantly carried about, and the great smoke-blackened gloomy house, produced an effect to which no description can do justice, and of which the sight of half-a-dozen Indians going through their dances for show, gives but a very faint idea. (*N* 280–82)

Despite my emphasis above on differences between Wallace's account of the savage and other roughly contemporaneous accounts, including especially Darwin's, this passage in many ways represents familiar discursive territory: that of the observer of "native" rituals recounting them with the language of the traveler on a quest for the new or bizarre. The specific content of what is witnessed is diminished or homogenized to the extent that it is rendered adjectivally as extraordinary, novel, wild, strange, curious. A closely related rhetoric characterizes the evocation of what appears to be the chaos of the feast, with its many participants involved in many different activities all at the same time; its unfamiliar noises and sights; its drinking. Bewilderment and confusion add to the novelty. Finally, the turn from such bewilderment to an invocation of the ineffable is itself predictable enough: the feast "produced an effect to which no description can do justice." But the end of the passage brings in another note, one that presages the account Wallace will give in his autobiography. For just as, in *My Life*, he establishes his long familiarity with what he calls the "tame Indians" of Brazil and Venezuela as that which was responsible for the extent of his surprise at the sight of "absolute uncontaminated savages," so here he introduces as the relevant prior experience a canned performance put on for tourists or metropolitan audiences: "half-a-dozen Indians going through their dances for show."

As his careful delineation of the appropriate contexts in which to understand his response to savages demonstrates, Wallace was acutely aware of the degree to which memories and expectations derived from prior experience or reading could give shape even to what are ostensibly first encounters. Although in this instance such experience seems only to have heightened his surprise, elsewhere it is held accountable for muted or delayed revelations. Whether in the form of the one or the other, the play of expectation met, disappointed, or exceeded is a staple of *A Narrative of Travels*. The entire book is framed in these terms insofar as the first chapter analyzes the initial effect of landfall in South America as occurring in the space between what Wallace expected and what he found: "The weather was not so hot, the people were not so peculiar, the vegetation was not so striking, as the glowing picture I had conjured up in my imagination, and had been brooding over during the tedium of a sea-voyage" (*N* 4). Wallace places half the blame for such a state of affairs on earlier travelers and writers who mislead by "crowd[ing] into one description all the wonders and novelties which it took them weeks and months to observe" (*N* 5).[29] But half the blame rests with the disappointed observer himself, who cannot see because he has not yet learned how to look. That observer, however, is not helpless to change the situation: Wallace details the possibility of an education of the eye, an education required before the wonders of the new become visible. "This is particularly the case," he adds, "with tropical countries" (*N* 4). Paradigmatic in this regard is his failure, during the first week in Brazil, to see a single parrot, hummingbird, or monkey. "And yet, as I afterwards found," he writes, these emblematic tropical creatures "are plentiful enough in the neighborhood of Pará; but they require looking for, and a certain amount of acquaintance with them is necessary in order to discover their haunts, and some practice is required to see them in the thick forest" (*N* 4).[30]

This "phenomenon of disappointment in relation to perception," argues Nancy Stepan in *Picturing Tropical Nature*, provides "insight into how fragile, even deceptive, popular representations of the world often are, and how they take cultural work to maintain."[31] Concerned specifically with popular representations of the tropics, Stepan reads *A Narrative of Travels* as "anti-tropical" to the extent that it "disrupts the [prevailing] conventions concerning the representation of tropical nature."[32] Wallace's disappointment on first seeing the tropics, she contends, evidences an ironic strain in his writing that works to complicate readerly assumptions about tropicality. Further, his insistence on familiarity and practice as the necessary precursors to accurate perception attests to a conviction that "seeing and representing … were culturally acquired skills."[33] The final clause of the description of the feast he witnessed on the Río Uaupés necessitates treating Wallace's descriptions of savages as analogous. Certain representations – in this case not the accounts of earlier travelers or the images produced by his own imagination but native dances put on "for show" – mislead, distorting the actuality by misrepresenting it. Wallace's task would then be a corrective one, and his description of savage life worthwhile specifically in that it possesses a high mimetic value. Ineffability itself constitutes part of that mimesis: evidently, to suggest the accuracy of one's own representation of a feast of the sort Wallace describes is to commit falsehood, for "no description can do justice" to it.

On this reading, Wallace's account in *A Narrative of Travels* of a "most extraordinary and novel scene" comports well with his assertion in *My*

Life that the "most unexpected sensation of surprise and delight" in his South American travels was encountering savages, and that "the surprise of it was that I did not in the least expect to be so surprised." The night spent in the malocca, Wallace suggests, shakes the complacency built up over several years of familiarity with indigenous peoples altered by contact with Europeans. But what, then, of the assertion made in the retrospective conclusion to *A Narrative of Travels* that I quoted earlier? "I do not remember a single circumstance in my travels," Wallace writes there, "that so well fulfilled all previous expectations, as my first view of the real uncivilized inhabitants of the river Uaupés." Far from being "unexpected" or a "surprise," in this rendering savages are that which most completely "fulfilled all previous expectations." For Wallace, it seems, savages are at once expected and unexpected, completely new and fully foreseen. One resolution to this apparent paradox is to be found in the possibility that savages fulfill expectations precisely to the extent that they exceed them. That is, if Wallace imagines savages to be so remarkable as to surpass what he can imagine or predict, they live up to what he expects only when they in fact prove to have been unimaginable and unforeseeable. Even given such a resolution, however, the tension between novelty and expectation in connection with savages persists, indicating the extent to which what Wallace found in the villages along the Uaupés affected him. Over the course of a long life he would often return in memory to this encounter. Amid widespread change and loss, it remained.

2

Writing up his South American voyage in the wake of the sinking of the *Helen* involved Wallace in the project of recreating fugitive memories using what materials he still possessed. Restorative and compensatory, such writing attempts to reproduce as text the experience that was to have been encoded in specimens. Thus *A Narrative of Travels* does a kind of double duty: it must represent the "beautiful forms from those wild regions" as well as the "recollections they would [have] call[ed] up" (*N* 401). The result, Wallace observes in the book's preface, leaves something to be desired:

I trust that the great loss of materials which I have suffered … may be taken into consideration, to explain the inequalities and imperfections of the narrative, and the meagreness of the other part of the work [i.e., the sections devoted to the natural history of the Amazon], so little proportionate to what might be expected from a four years' residence in such an interesting and little-known country. (*N* iv)

Darwin apparently did not take these conditions of production into consideration. In a letter to Henry Walter Bates dated December 3, 1861, he complains: "I was a *little* disappointed in Wallace's book on the Amazon; hardly facts enough."[34] Text may stand in for absent specimens and memories, but only imperfectly and meagerly; evidence of "great loss of materials" persists.

Given the context of such loss, we might speculate that it was in pursuit of further and more successful restoration that Wallace, after a stay in England of less than two years and despite vowing "never to trust [himself] more on the ocean," departs once again – this time for a several-thousand-mile-long odyssey among the chain of islands that constitute present-day Malaysia and Indonesia (*ML* 1: 309).[35] In the preface to the book documenting that journey, *The Malay Archipelago: The Land of the Orang-Utan, and the Bird of Paradise; A Narrative of Travel, with Studies of Man and Nature* (1869), Wallace counters the insufficiencies mentioned in the prefatory remarks to *A Narrative of Travels* by enumerating the astonishing number of specimens collected and safely conveyed to England over the course of his eight years in Southeast Asia: 310 mammals, 100 reptiles, 8,050 birds, 7,500 shells, 13,100 butterflies, 83,200 beetles, and 13,400 other insects – all told, "125,660 specimens of natural history" (*MA* 1: xiv). Making up for South American collections lost and then some, Wallace builds an archive of archipelagic recollections second to none.

But the frequent experience or prediction of loss characterizes the Malay travels as well. While nothing that befalls Wallace quite equals the catastrophe of fire at sea, myriad smaller troubles plague him. Injuries, false starts, delays, and lost opportunities abound. Having wounded his ankle not long after arriving in New Guinea, for instance, he is forced to remain indoors until it heals. Just as, in describing the *Helen* incident, Wallace gives short shrift to his corporeal suffering, focusing rather on lost specimens, so in the account of this injury he minimizes attention to the injury itself and dwells instead on its consequences for his work in natural history. After a brief summary of the course of treatment followed over the several weeks it took him to recover, he turns to what he takes to be the real hardship: missing out on the chance to canvass his surroundings. "And this, too," he exclaims,

in New Guinea! A country which I might never visit again, – a country which no naturalist had ever resided in before, – a country which contained more strange and new and beautiful natural objects than any other part of the globe. The naturalist will be able to appreciate my feelings, sitting from morning to night in my little hut, unable to move without a crutch. (*MA* 11: 316)[36]

The would-be collector injured and immobile, incapable of taking advantage of the opportunity to discover an unknown country literally out-side his door: such a self-portrait recalls, in addition to the shipwreck pas-sage of *A Narrative of Travels*, the sustained attention paid in that book to the interplay of expectation, experience, and disappointment. So, too, does much else in *The Malay Archipelago*. Wallace once more takes pains to contradict what he terms the "usual ideas of the vegetation of the tropics," which, with their promise of breathtakingly colorful floral profusion, inev-itably lead to a feeling of anticlimax in the face of an actual tropical scene (*MA* II: 294). Again faulting those travel writers whose desire to impress comes at the expense of fidelity to nature, he adds the effect of visits to European "hothouses" and "flower-shows," where "we gather together the finest flowering plants from the most distant regions of the earth, and exhibit them in a proximity to each other which never occurs in nature" (*MA* II: 297).[37] It seems a matter of course, then, that of the first sight of an active volcano – glimpsed from the deck of a Bugis prau while passing the Banda Islands on his journey from Celebes (now Sulawesi) to the Aru Islands in search of birds of paradise – Wallace observes: "but pictures and panoramas have so impressed such things on one's mind, that when we at length behold them they seem nothing extraordinary" (*MA* II: 173). The visitor from the European metropolis, having read about tropical flora and volcanic activity in books, visited hothouses, and seen panoramas, is inured to the sight of these wonders long before actually encountering them.[38]

Given such preconditioning, the traveler, to see at all, must learn to see again. In *A Narrative of Travels* such a process of relearning depends on patient receptiveness to the realities of the new. In *The Malay Archipelago*, that receptiveness itself is shown to be contingent on a form of unlearning or defamiliarization.[39] Thus elsewhere in the book Wallace writes of the same volcano that had at first failed to seem extraordinary:

It is only when actually gazing on an active volcano that one can fully realize its awfulness and grandeur. Whence comes that inexhaustible fire … ? Whence the mighty forces that produced that peak … ? The knowledge from childhood, of the fact that volcanoes and earthquakes exist, has taken away somewhat of the strange and exceptional character that really belongs to them. The inhabitant of most parts of northern Europe, sees in the earth the emblem of stability and repose. His whole life-experience, and that of all his age and generation, teaches him that the earth is solid and firm … A volcano is a fact opposed to all this mass of experience, a fact of so awful a character that, if it were the rule instead of the exception, it would make the earth uninhabitable. (*MA* I: 450–51)[40]

As with the biota of South America, so with that of Southeast Asia, and so with volcanoes: familiarity based on representations or recreations (in books, hothouses, panoramas, "native" shows) impedes true vision; the thing itself, approached rightly, restores that vision. A volcano may "seem nothing extraordinary" and yet, properly understood, constitutes a "fact" strange, exceptional, and awful. Such correct understanding requires direct experience ("It is only when actually gazing on an active volcano that one can fully realize its awfulness and grandeur"), but that experience alone is not sufficient. It must be supplemented with meditation or ratiocination such that the implications of the "fact" it represents come to light – especially when, as in this case, that fact stands as a contradiction of the viewer's "knowledge from childhood" and "mass of experience." What one remembers, Wallace suggests, sometimes obscures what one sees; true seeing and knowing may require forgetting.

I take such pains to establish the pervasive presence in *The Malay Archipelago* of what Wallace presents as a dynamic among preconceptions, encounters with the empirical, and the work of forgetting required to realize the meaning of such encounters because that dynamic organizes the book's sustained engagement with the question of savages. Making good on the promise in its subtitle that *The Malay Archipelago* will feature "Studies of Man" as well as of "Nature," Wallace begins and ends the book with humans: its celebrated frontispiece pictures an "Orang Utan Attacked by Dyaks," while its final chapter treats "The Races of Man in the Malay Archipelago."[41] In between, Wallace maintains a running commentary on the myriad sorts of peoples with whom he comes into contact. Of that commentary's many preoccupations, I will focus on the two that are of most consequence in Wallace's memory of savages and the effect of that memory on his conclusions about evolution and the relations between savage and civilized. In connection with the first such preoccupation, colonialism, Wallace vacillates between endorsing and questioning the colonial enterprise and the racial hierarchy that provides one of its chief rationales. In connection with the second, human progress, he also vacillates but finally announces as indisputable the superiority of the savage over the civilized in all areas of life except the material or technological. Thus *The Malay Archipelago* is a text riven with contradictions – contradictions that derive, in part, from Wallace's memories of his first encounter with "absolute uncontaminated savages" and that look forward to his belief that even such savages are so fully human as to render impossible their descent from other animals by natural causes alone.

Most of the areas through which Wallace passes during his Malay travels were under occupation by the Dutch or the British. As a result, the question of colonialism looms large in his thinking about humans in *The Malay*

Archipelago. The whole colonial enterprise, for Wallace, remains to be properly thought through: "The true 'political economy' of a higher, when governing a lower race, has never yet been worked out. The application of our 'political economy' to such cases invariably results in the extinction or degradation of the lower race" (*MA* 1: 456). The reference to "our 'political economy'" is to British colonialism, and this estimate of the deleterious effects of Britain's rule on "lower race[s]" is typical of Wallace in his Malay book. Recalling the prediction about the fate of Amazonian tribespeople with which he closes *A Narrative of Travels*, such an estimate places the British Empire on par with "the refuse of Brazilian society": for uncontaminated savages, contact with either is likely to result in destruction (*N* 519).

Wallace's critique of British colonialism derives, as he sees it, from his presence on the ground. As such, it constitutes one aspect of what he views as the corrective force of that presence – like his realization of the "fact" of a volcano, one form of relearning to which his travels have contributed. But such relearning does not, apparently, contradict in its entirety his "mass of experience," for Wallace does not, in *The Malay Archipelago*, come to oppose colonialism *tout court*. He regularly singles out the British for mistaken ideas about colonial governance but just as regularly lauds the Dutch for their correct ones. Furthermore, even if the most successful or appropriate form of colonial rule has yet to be ascertained to Wallace's satisfaction, he seems to take for granted that some races will rule others. Indeed, the Dutch are praised in particular for understanding that "higher" and "lower" "races" exist and have different needs, and for tailoring their colonial policy accordingly. Thus, although English observers often fault the Dutch for what they take to be their despotic treatment of colonial subjects, "[c]hildren," Wallace writes, "must be subjected to some degree of authority, and guidance" – and "there is not merely an analogy, – there is in many respects an identity of relation, between master and pupil or parent and child on the one hand, and an uncivilized race and its civilized rulers on the other" (*MA* 1: 398). "There are," he continues, "certain stages through which society must pass in its onward march from barbarism to civilization" (*MA* 1: 402). To attempt to skip a stage – to invite savages to participate in a democratic process of self-governance, for instance, before submitting them to a period of despotism – is to misapprehend their nature and, in so doing, to subject them to misrule. Wallace was eventually to reject such ideas and to insist that some form of democratic self-rule, however minor and gradually introduced, was the only appropriate form of government for colonized peoples.[42] But the stadialism and paternalism largely missing from

Wallace's account of South American indigenes in *A Narrative of Travels* appear in *The Malay Archipelago* in unmitigated forms.[43]

In relation to the dynamic between preconceived notions and encounters with the empirical, it might be said that Wallace's Malay travels dislodge certain nationally specific ideas of what constitutes good colonial governance only to confirm other, more general assumptions subtending the practice of colonialism as such: namely, the existence of a hierarchy of "races"; Europeans' status in that hierarchy as higher or more mature than other, "lower" or more childlike races; and thus, finally, the duty of the former to "civilize" the latter.[44] But there is for Wallace a second moment in the process of unlearning, a moment when this knowledge, too, is shaken by experience and shown to be in need of being forgotten. Immediately following the sentence detailing the "analogy" or "identity of relation" between parent and child and civilized and savage, Wallace observes: "We know (or think we know) that the education and industry, and the common usages of civilized man, are superior to those of savage life; and, as he becomes acquainted with them, the savage himself admits this" (*MA* I: 399). What, one wonders, is the effect of that parenthetical qualification? What force can the offhand "or think we know" have in the midst of such otherwise confident pronouncements about "civilized" peoples, "savages," and the relations between them? Is it plausible to imagine that the "think we know" casts doubt on such knowledge, opening up the possibility that, like knowledge of tropical vegetation, it, too, has resulted from misrepresentation and leads to misperception?

An affirmative answer is suggested by the fact that other strains of thinking about savages in *The Malay Archipelago* play on and develop this doubt. Thus, for instance, Wallace makes frequent observations about the contrast between what one expects of a savage or understands savagery to be and the ways of savages themselves. Like tropical vegetation, like volcanoes, savages have been misrepresented, and perceiving them aright requires direct experience as well as intellectual effort on the part of the observer to overcome or discard preconceptions. So, for instance, he comments on the inhabitants of Dorey in western New Guinea (now the Indonesian province of Irian Jaya):

They live in the most miserable, crazy, and filthy hovels, which are utterly destitute of anything that can be called furniture … Their food is almost wholly roots and vegetables … and they are consequently very subject to various skin diseases, the children especially being often miserable-looking objects, blotched all over with eruptions and sores. If these people are not savages, where shall we find any? Yet they have all a decided love for the fine arts, and spend their leisure time in executing works whose good taste and elegance would often be admired in our schools of design! (*MA* II: 325)

The surprise, signaled by yet another of Wallace's beloved exclamation points, inheres in the apparent difference between what is understood about or expected of savagery ("If these people are not savages, where shall we find any?") and the artistic production of a particular group of savages. The "yet" both points to a contrast between the two and offers the "good taste and elegance" of the art produced by the people of Dorey as giving the lie to that contrast. Rhetorically, these assertions take the shape of a distorted syllogism, one supplemented by a conclusion that throws what precedes it into doubt. Such a syllogism might read: savages live and eat poorly; the people of Dorey live in "filthy hovels" and eat only "roots and vegetables"; therefore, the people of Dorey are savages; but these people produce art worthy of admiration, and thus they are also civilized.[45] In a gesture representative of his thought processes, Wallace pushes past that which is confidently "known" to arrive at puzzlement or skepticism.

What in this account takes the shape of doubt as to the validity of the category "savage" at other moments assumes the more explicit form of a reversal of savage and civilized. After being informed by Aru islanders among whom he was staying, for instance, that visitors from all around were making pilgrimages to see him, "the first … real white man [who] had come among them," Wallace observes: "A few years before I had been one of the gazers at the Zoolus and the Aztecs in London. Now the tables were turned upon me, for I was to these people a new and strange variety of man, and had the honour of affording to them, in my own person, an attractive exhibition, gratis" (*MA* II: 241). The concluding pronouncement is sharper in Wallace's journal entry for April 6, 1857: "I was to the Arru [*sic*] Islanders a new & strange variety of man, & had the pleasure of affording them in my own person an instructive lesson in comparative Ethnology."[46] As with the phrase "or think we know" in "We know (or think we know) that the education and industry, and the common usages of civilized man, are superior to those of savage life," it is difficult to parse the precise sense of these observations. It is perhaps too easy to view them as instances of cultural relativism inhering in the reversal of the observer/observed dichotomy. In fact, the late tonal shift in the direction of sarcasm, especially marked in the final "attractive exhibition, gratis," may invite an evaluation of the passage as confirming rather than undermining Wallacean (and hence British and European) superiority. Reading this instance of reversal as merely effecting such confirmation, however, would itself underplay the sense of displacement to which these passages give voice, sarcastic or not – a displacement familiar not only from other passages in this book but also from Wallace's initial encounter with savages as narrated in *A Narrative of Travels*: "I felt that I was as much in the midst of something new and startling, as if I had been instantaneously transported to a distant and unknown country" (*N* 477). Here and elsewhere in *The Malay Archipelago*, then, Wallace evinces a doubleness in relation to savages. On one hand, he seems content to rest assured in the knowledge of their childlike status, their "lower" position in a developmental scheme that culminates in Europeans such as himself. On the other, he dwells on disruptions to such a scheme: doubt as to the superiority of European civilization, evidence of savage art as tasteful as any produced in British design schools, savage interest in human difference of a sort that mirrors that of the naturalist in pursuit of scientific knowledge.[47]

Such doubleness reaches its apogee in the final chapter, "The Races of Man in the Malay Archipelago." The whole chapter is organized around a confident racial typology: "Two very strongly contrasted races," begins the chapter's second paragraph, "inhabit the Archipelago – the Malays … and the Papuans" (*MA* II: 439). Indeed, so certain of the accuracy of this typology is Wallace that he considers it as part of the evidence for the existence of what has come to be known as Wallace's Line, a division that sharply demarcates the groups of flora and fauna in the islands among which he traveled and, in so doing, reveals those islands' history.[48] The human version of this line, avers Wallace, "separate[s] the Malayan and all the Asiatic races, from the Papuans and all that inhabit the Pacific; and though along the line of junction intermigration and commixture have taken place, yet the division is on the whole almost as well defined and strongly contrasted, as is the corresponding zoological division of the Archipelago, into an Indo-Malayan and Austro-Malayan region" (*MA* II: 452–53). As his names for these regions imply, Wallace concludes that the division he believes he has discovered admits of a clear geological interpretation: that the western or "Indo-Malayan" part of the archipelago once formed part of the Asian continental landmass, while the eastern or "Austro-Malayan" part was once closely associated with Australia. The distribution of the two "strongly contrasted races" mirrors the distribution of plants and animals closely enough to serve as another confirmation of the accuracy of that conclusion.

The account of the two different "races" in the islands and what Wallace views as their different abilities, propensities, and limitations also leads to pronouncements on their likely fate under colonial rule. The adaptable

"Malay race," being "the most civilized," will thrive, but "there can be little doubt of the early extinction of the Papuan race. A warlike and energetic people, who will not submit to national slavery or to domestic servitude, must disappear before the white man as surely as do the wolf and the tiger" (*M* II: 439, 458). The approbation implied in the descriptors "warlike and energetic" as well as in the contention that Papuans will brook no "slavery" or "servitude" is immediately rendered moot by the prediction of their certain disappearance. Moreover, the comparison of these people with wolves and tigers bestializes them at the same moment that it naturalizes their extinction.

Near the end of the final chapter of *The Malay Archipelago*, however, in a series of astonishing paragraphs that follow a break inserted after the sentence just quoted, Wallace looks back at savages and, in doing so, recalls the various moments where they have surprised or impressed him. The context is the familiar one of human development or progress, but the vantage from which such progress is considered is its putative end-point: what Wallace describes as the "ideally perfect social state." This peroration is well known, but I quote from it at length nonetheless, both because of its intrinsic interest and because the specific context of the discussion of savages in *A Narrative of Travels* and *The Malay Archipelago* provides powerful critical purchase on its animadversions. Wallace begins thus:

> Before bidding my readers farewell, I wish to make a few observations on a subject of yet higher interest and deeper importance, which the contemplation of savage life has suggested, and on which I believe that the civilized can learn something from the savage man.
>
> We most of us believe that we, the higher races, have progressed and are progressing. If so, there must be some state of perfection, some ultimate goal, which we may never reach, but to which all true progress must bring us nearer. What is this ideally perfect social state towards which mankind has ever been, and still is tending? Our best thinkers maintain that it is a state of individual freedom and self-government, rendered possible by the equal development and just balance of the intellectual, moral, and physical parts of our nature, – a state in which we shall each be so perfectly fitted for a social existence, by knowing what is right, and at the same time feeling an irresistible impulse to do what we know to be right, that all laws and all punishments shall be unnecessary. In such a state every man would have a sufficiently well-balanced intellectual organization to understand the moral law in all its details, and would require no other motive but the free impulses of his own nature to obey that law. (*MA* II: 459–60)

Here, at the end of a chapter in which Wallace has dispassionately and confidently adjudicated among differing savage types with the same taxonomizing impulse with which he adjudicates among species of beetles, he turns the tables once again. Having praised the Dutch for understanding that savages must be treated by their European governors as if they were children or pupils, he inverts that relation, insisting, in connection with this "subject of yet higher interest and deeper importance" than any treated earlier in *The Malay Archipelago*, that "the civilized can learn something from the savage man." That subject, improbable as it seems, is the same as that which the Dutch seem to have mastered: progress. But as with the earlier caveat, "or think we know," only more forcefully, what had heretofore been thought about progress is cast into doubt.

"We most of us believe that we, the higher races, have progressed and are progressing," Wallace begins, and already the suggestion of complacency or potentially mistaken belief stands out. But the question of the soundness or unsoundness of the notion that "the higher races ... are progressing" is momentarily left behind in favor of consideration of what such progress, if taking place at all, progresses toward – what its goal or end-point might be. Invoking the opinion of "[o]ur best thinkers," Wallace submits that the goal must be considered to be a society in which "every man would have a sufficiently well-balanced intellectual organization to understand the moral law in all its details, and would require no other motive but the free impulses of his own nature to obey that law." In short, if not quite Friedrich Engels's "withering away of the state," Wallace posits as the goal of the progress of civilization a withering away of the juridical – a rendering supererogatory of explicitly codified law and specifically of law enforcement because of an internalization, indeed a naturalization of the law.[49] "[T]he moral law," in a perfect social state, would not only be understood by each individual, but would in fact be indistinguishable from no law at all – would correspond with "the free impulses of [each person's] own nature."

Already, then, the perfect social state, the goal of progress, has in essence been redefined as a particular state of nature. So to the attentive reader it may not come as an entire surprise when Wallace continues as follows:

Now it is very remarkable, that among people in a very low stage of civilization, we find some approach to such a perfect social state. I have lived with communities of savages in South America and in the East, who have no laws or law courts but the public opinion of the village freely expressed. Each man scrupulously respects the rights of his fellow, and any infraction of those rights rarely or never takes place. In such a community, all are nearly equal. There are none of those wide distinctions, of education and ignorance, wealth and poverty, master and servant, which are the product of our civilization; there is none of that wide-spread division of labour, which, while it increases wealth, produces also conflicting interests; there is not that severe competition and struggle for existence, or for wealth, which the dense population of civilized countries inevitably creates ...

Now, although we have progressed vastly beyond the savage state in intellectual achievements, we have not advanced equally in morals ... [I]t is not too much to say, that the mass of our populations have not at all advanced beyond the savage code of morals, and have in many cases sunk below it ...

Compared with our wondrous progress in physical science and its practical applications, our system of government, of administering justice, of national education, and our whole social and moral organization, remains in a state of barbarism. (*MA* ii: 460–62)[50]

Building on earlier moments of doubt about the superiority of civilized to savage (we "think we know"), where the European observer sees himself as observed, where savages produce admirable art, this is reversal writ large. The notion of progress and development inherent in stadialism and so often invoked to demonize non-European peoples here turns back on itself and provides the basis for a critique of "civilization." Wallace achieves such a critique in part by distinguishing among different kinds of progress, so that while in "physical science" and "its practical applications" the so-called civilized undoubtedly excel, in all that remains – government, education, and "social and moral organization" in its entirety – some savages stand closer to perfection, for they possess that internal "moral law" toward which the civilized strive.[51] But, as the beginning of this series of paragraphs makes clear, for Wallace advancement in technological prowess and scientific knowledge can only be a secondary goal. "[A]ll true progress" is progress toward a just social state, and, measured against savages in relation to that goal, the civilized come up short.

Recollection, and specifically recollection of savages, enables this sweeping attack on civilization, an attack that co-opts stadialism in the service of inverting its underlying assumptions and explicit ideological investments. Fusing his memories of the people of the Río Uaupés with those of his Malay travels and bringing both to bear on the question of progress, Wallace writes: "I have lived with communities of savages in South America and in the East." All that follows derives from this synthetic retrospect. Revisiting and redeploying the account of domestic intimacy with savages first provided in *A Narrative of Travels*, Wallace offers up his memories of that intimacy as providing evidence that defamiliarizes not simply preconceptions about savages themselves but logical and temporal relations between savagery and civilization. These recollections of encounters with "barbarism" reveal, startlingly, that the term must properly be used to signify the opposite of its conventional meaning. Thus savagery and barbarism, with their connotations of filth, ugliness, brutishness, and the primitive, are detached from "savages" themselves and associated with what had been understood to have outgrown them: the civilized.[52] In a moment at once as close to and as far from Darwin's account of Fuegians as he will ever come, Wallace embraces the notion that savages resemble prehistoric ancestors even as he asserts that civilized Europeans have degenerated in relation to such ancestors, who themselves represent the telos of civilization.[53] Suddenly taking the shape of living mnemonic devices, Wallace's savages provide access to what can only be described as a memory of the future.

3

Throughout this chapter I have sought to demonstrate what made savages unforgettable for Wallace and to trace some of the consequences of that unforgettability. Examining the moments in which savages appear in *A Narrative of Travels*, *The Malay Archipelago*, and *My Life*, I have argued that Wallace remembered savages because of the way they simultaneously confirmed and exceeded his expectations. So startling that they required he relearn to see, forgetting all he thought he knew about them, savages were also so familiar as to confirm the existence of a world wholly distinct from that of "civilized" Europe – a world whose enviably harmonious social arrangements Wallace was, for a time, able to share. These memorable aspects of savagery, I have further argued, eventuate in a series of closely related reversals: local reversals of perspective whereby Wallace the observer finds himself the observed as well as global reversals of definition and value whereby savages come to exemplify perfection and progress while civilization incarnates barbarism. It remains to explore the decisive role of savages in what was, for his fellow biologists and for his conception of the human, Wallace's most momentous reversal of all. In an 1869 review, "Sir Charles Lyell on Geological Climates and the Origin of Species," and then in a subsequent essay expanding on the claims made in that review, "The Limits of Natural Selection as Applied to Man," the concluding chapter of his *Contributions to the Theory of Natural Selection* (1870), Wallace announced that he had been mistaken about humans. Although demonstrably the product of evolutionary processes, humans cannot, he claims, have resulted from those processes alone. Supernatural guidance must have been necessary, and what he recalls about savages proves it.

In "Sir Charles Lyell on Geological Climates and the Origin of Species," Wallace takes the 1868 publication of the tenth edition of *Principles of Geology* as an occasion for a largely hagiographic treatment of Lyell and his work. He reminds readers of the dominance of catastrophist geology at the beginning of the nineteenth century and the eclipse of its influence with the 1830 appearance of the first volume of the first edition of *Principles* ("Lyell," 364). Rehearsing Lyell's arguments for uniformitarianism, he insists on the sufficiency of past time for the formation of the present state of the earth by uniformitarian means. He also treats glaciation, speculating on the phenomena that most likely cause periodic worldwide cooling trends, and considers the question of the imperfection of the fossil record, repeating Lyell's and Darwin's analogy between that record and a book extant only in small part: "With this mass of evidence, all tending in one direction, we can scarcely deny the fairness of Mr. Darwin's assumption that not only is the geological record imperfect, but we only have fragments of the last volume of it: an unknown series of preceding volumes having been lost or destroyed" ("Lyell," 389–90). The history of the earth converges with Wallace's own history after the *Helen*: both comprise loss and destruction and so both must be reconstructed by means of prosthetic substitutes for what can no longer be read or remembered.

If there is much of interest here, there is little that surprises. In the review's final few pages, however, what had been an instructive if quotidian document in the workings of what Thomas Kuhn called normal science takes a sharp turn as Wallace considers the implications for understanding humans of what is, in his view, "the great distinguishing feature of this edition [of *Principles*], – the adoption in its main outlines, if not in all its details, of Mr. Darwin's theory of the Origin of Species" ("Lyell," 379).[54] Chief among those implications, Wallace points out, is "the logical necessity" of "the derivation of man from the lower animals," and on this much he agrees with Lyell and Darwin ("Lyell," 391). But he differs in his understanding of how this derivation came about, and in entering into the details of that difference defects spectacularly from the camp of biologists committed to a strictly naturalistic explanation of all the phenomena of life – a camp of which, since the publication of "On the Tendency of Varieties to depart indefinitely from the Original Type" in 1858, he had been one of the staunchest members.

Convinced of the descent of humans from other animals, Wallace nonetheless also believes, he now announces, that "[man's] intellectual capacities and his moral nature were not wholly developed by the same process" as that which gave rise to those animals – that is, by unaided natural selection. "[T]he moral and higher intellectual nature of man," he avers, is a "unique phenomenon" in the natural world. Further, even the human body features "unique" traits "not explicable on the theory of variation and survival of the fittest." These include, Wallace claims, the "brain, the organs of speech, the hand, and the external form" ("Lyell," 391). Recanting his earlier position, flying in the face of Darwinian orthodoxy, and abandoning the criterion of naturalistic explanation, Wallace announces that such human characteristics, because inexplicable on the theory of evolution by natural selection alone, attest to the presence of some higher power that has guided evolution in its path toward the human. "While admitting to the full extent the agency of the same great laws of organic development in the origin of the human race as in the origin of all organized beings," he writes, "there yet seems to be evidence of a Power which has guided the action of those laws in definite directions for special ends" ("Lyell," 393). Darwin, in a letter to Lyell, responded with dismay and disbelief: "What a good sketch of natural selection! but I was dreadfully disappointed about Man, it seems to me incredibly strange … and had I not known to the contrary, would have sworn it had been inserted by some other hand."[55]

Just as they prove the barbarity of civilization, savages for Wallace evince the failure of evolutionary theory to account for the human. Different as the two conclusions are, both depend on a conviction, itself contrary to much contemporary scientific as well as lay opinion, and specifically at odds with the comparative method, as to the essential unity of the human across not only space but also time. In "Sir Charles Lyell," that unity is shown in the first instance by a comparison of savage and prehistoric with civilized brains: "In the brain of the lowest savages, and, as far as we yet know, of the pre-historic races, we have an organ … little inferior in size and complexity to that of the highest types (such as the average European)" ("Lyell," 391). As in *The Malay Archipelago*, Wallace here trades in the familiar racist typology of "lowest" and "highest" examples of humanity, examples that correspond to savages and Europeans, respectively. But, again as in *The Malay Archipelago*, such a typology is invoked only to be undone, for the brain of the savage and of the prehistoric human is "little inferior" to that of the European. If the brains of savages and Europeans differ little, however, their "mental requirements," Wallace claims, differ enormously, and it is specifically the perceived mismatch between mental capacity and mental need that leads to doubt as to the possibility that natural selection gave rise to those brains: "Natural selection could only have endowed the savage with a brain a little superior to that of an ape, whereas he actually possesses one but very little inferior to that of the average members of our learned societies" ("Lyell," 392).

Memory plays a palpable but muted role here insofar as, although he makes no explicit claims to this effect, Wallace's notions about the intellectual demands of savage life derive from his experience in South America and Southeast Asia. In the next stage of the argument, however, recollection figures centrally. "Those who have lived much among savages know," Wallace writes, "that even the lowest races of mankind, if healthy and well fed, exhibit the human form in its complete symmetry and perfection" ("Lyell," 392).[56] In a formulation echoing that found at the end of *The Malay Archipelago* ("I have lived with communities of savages in South America and in the East"), Wallace calls on personal experience to argue that savages (and, by extension, prehistoric humans as well) are as physically perfect as modern Europeans. Recalling his time among indigenous peoples in South America and Southeast Asia, Wallace places it in evidence to give the lie to the notion not only of mental but also of physical advance from savage to civilized.[57] This claim is more momentous than it might at first seem, for Wallace finds in humans' physical characteristics the wellspring of their "intellectual capacities" and "moral nature." "The supreme beauty of our form and countenance," he argues, "has probably been the source or all our aesthetic ideas and emotions." Moreover, "our naked skin, necessitating the use of clothing, has at once stimulated our intellect, and by developing the feeling of personal modesty may have profoundly affected our moral nature" ("Lyell," 393).

From memory and study Wallace concludes that all known humans, past and present, evidence physical perfection – perfection that puts the role of natural selection in producing humans in doubt because it admits of no gradation and so contradicts an inescapable corollary of that selection. Further, again in contradiction to what natural selection requires, humans' physical perfection also has a telos insofar as it seems to have been calculated to spur on the development of what are regarded as still more characteristically human qualities: intellect and morality. Thus, writes Wallace, in the final sentence of the review: "[L]et us not shut our eyes to the evidence that an Overruling Intelligence has watched over the action of those laws [the laws of natural selection], so directing variations and so determining their accumulation, as finally to produce an organization sufficiently perfect to admit of, and even to aid in, the indefinite advancement of our mental and moral nature" ("Lyell," 394).

Wallace repeats and expands on the assertions put forth at the end of his review of Lyell's *Principles* in "The Limits of Natural Selection as Applied to Man." Refining his argument and drawing out its implications, he now contends that natural selection fails to account for not one but two milestones in the history of life: the appearance of human beings but also, before that, the origin of consciousness. Consciousness is treated in the airily metaphysical final section of the essay, which argues against the supposition that it results from a certain organization of matter. Largely via an attack on Huxley's "On the Physical Basis of Life" (1868), Wallace contends that there is no matter but only "force," that all force is "will force," and that both consciousness and humans are the result of unknown powers giving shape to the evolutionary process in exactly the same way humans themselves have shaped that process in their efforts to domesticate plants and animals.[58] But leading up to this discussion of consciousness is, once more, a meditation on savages and what they show about the human.

As in the review of *Principles*, Wallace considers first the savage brain. And, as we have seen so often, he insists that proper consideration of that brain forces a re-evaluation of common assumptions. Thus he opens with a discussion of craniometry, the racialist science that sought to establish a correlation between the size of the brain and intellectual ability and, further, to establish a hierarchy of races based on cranial capacity. As Wallace notes, "all the most eminent modern writers see an intimate connection between the diminished size of the brain in the lower races of mankind, and their intellectual inferiority."[59] But Wallace immediately puts into evidence a collection of brain volume numbers that go to prove "that the absolute bulk of the brain is not necessarily much less in savage than in civilised man" ("Limits," 336).[60] Leaving unquestioned the correlation between the size of the brain and intellectual ability, Wallace contests instead the claim that size differs significantly among the "races." Further, turning next to the fossil record, he contends: "what is still more extraordinary, the few remains yet known of pre-historic man do not indicate any material diminution in the size of the brain case" when measured against contemporary humans, whether "savage" or "civilized" ("Limits," 336). The conclusion he draws is identical to that in his review of the tenth edition of Lyell's *Principles*: savages and prehistoric humans have more intellectual capacity than required for survival, and such excess capacity violates the stipulation that natural selection can only preserve variations immediately useful to their possessor. "[W]e cannot fail to be struck," he writes, "with the apparent anomaly, that many of the lowest savages should have as much brains as average Europeans. The idea is suggested of a surplussage of power; of an instrument beyond the needs of its possessor" ("Limits," 338).

The evidence of craniometry alone, however, although apparently convincing, is not all that Wallace brings to bear on the question. After enumerating skull sizes, Wallace considers the "moral and aesthetic faculties" of the savage, claiming "Any considerable development of these would … be useless or even hurtful to him, since they would to some extent interfere with the supremacy of those perceptive and animal faculties on which his very existence often depends" ("Limits," 340). "Yet," he continues, "the rudiments of all these powers and feelings undoubtedly exist in him" ("Limits," 341). In fact, "Instances of unselfish love, of true gratitude, and of deep religious feeling, sometimes occur among savage races" ("Limits," 341). Drawing on his wide reading in ethnography as well as "[his] own experience among savages," Wallace claims that, although at best unneeded and at worst actually harmful to survival in a state of nature, benevolence, cooperation, and the love of truth undoubtedly exist among indigenous peoples ("Limits," 354). As with the size of the brain, these data contradict the dictates of natural selection and so militate in favor of its selective use by a power beyond nature: "The inference I would draw from this class of phenomena is, that a superior intelligence has guided the development of man in a definite direction, and for a special purpose, just as man guides the development of many animal and vegetable forms" ("Limits," 359). Exploiting the implication of a guiding consciousness inherent in the metaphor of "natural selection," Wallace reverses the direction of Darwin's argument in the *Origin*, which arrives at natural selection by way of an extended treatment of human or artificial selection, in order to give a kind of scientific credibility to his spiritualist conclusions about human evolution. Denying that his conclusions about humans contradict the theory of evolution by means of natural selection, he argues:

I do not see that the law of "natural selection" can be said to be disproved, if it can be shown that man does not owe his entire physical and mental development to its unaided action, any more than it is disproved by the existence of the poodle or the pouter pigeon, the production of which may have been equally beyond its undirected power. ("Limits," 370)

Although Wallace was to live and work into the twentieth century, these essays of 1869 and 1870 show the consequences of his encounters with and memories of savages at their most far reaching. For in them Wallace calls upon those encounters and memories to disprove his own earlier position on human evolution. Like Darwin but with far different results, Wallace recalls the savage as an exception. Darwin's memory of Fuegians sets them apart from both other animals and other humans, sacrificing them in the interest of securing an unbroken continuum of life, joined as a family is joined both through time and in space, related but not the same. Wallace's memories of South American and Southeast Asian indigenes lead to the conclusion that not they alone but rather humans as such constitute an exception. Positing a break in the continuum of living beings as well as in the nature of the forces at work in their production, Wallace's savage mnemonics insist on recognition of the essential oneness of the human – and hence, for Wallace, its supernatural provenance.

4

What, in addition to savages, does Wallace find memorable? Near the mid-point of *The Malay Archipelago*, he describes living through an earthquake as giving rise to a "sensation" that is "never to be forgotten":

We feel ourselves in the grasp of a power to which the wildest fury of the winds and waves are as nothing; yet the effect is more a thrill of awe than the terror which the more boisterous war of the elements produces. There is a mystery and an uncertainty as to the amount of danger we incur, which gives greater play to the imagination, and to the influences of hope and fear. These remarks apply only to a moderate earthquake. A severe one is the most destructive and the most horrible catastrophe to which human beings can be exposed. (*MA* I: 394)

The sentiments of the passage closely resemble Wallace's reflections on looking at an active volcano. Just as a volcano, correctly understood, contradicts all that Europeans assume about the stability of the ground beneath their feet, so, too, does an earthquake. If there is a significant difference between the two passages it is that, unlike in the case of the volcano, little ratiocination seems necessary to realize what an earthquake portends. Its tangible power, more forceful even than that of a storm at sea, makes an immediate impression. So, on one hand, the observer (or, better, "experiencer" – since Wallace's position seems to lack the distance necessary for observation) need bring little to bear on an earthquake in order to realize it and its implications in their fullness. But, on the other, because of the "mystery and … uncertainty" as to just how imminent a threat to life the earthquake poses, the imagination proves crucial. Producing the same contradiction to the "mass of experience" of a stable earth as a volcano does, the earthquake differs insofar as something of that contradiction is immediately felt. Despite Wallace's constant refrain about the need for learning to look and feel anew before new things can be appreciated or even properly perceived, to forget in the service of making a new and better memory, the earthquake insists on its own novelty. Yet something, too, remains to be imagined, hoped, and feared. We might call an earthquake as Wallace describes it sublime, with the proviso that in this instance of sublimity some actual threat of death is apparently present. But how much of a threat, what specifically to fear, what to hope – so much is left in doubt. Enormous power, the shaking of what had seemed solid, a danger to one's existence but a danger the precise extent of which must be imagined: all this, taken together, makes an earthquake memorable, "never to be forgotten."

As such a description should make evident, for Wallace encountering savages was like living through an earthquake. In the final chapter of *A Narrative of Travels*, Wallace pens a sentence I have already quoted on more than one occasion: "I felt that I was as much in the midst of something new and startling, as if I had been instantaneously transported to a distant and unknown country" (*N* 477). That sentence repeats an earlier one: "It was the men however who presented the most novel appearance, as different from all the half-civilized races among whom I had been so long living, as they could be if I had been suddenly transported to another quarter of the globe" (*N* 277). In both cases, looked at again in light of the passages on earthquakes and volcanoes, what had once appeared to indicate a feeling of physical displacement now reads as a description of the earth moving beneath Wallace's feet. There is a vertiginous sense of what had seemed solid suddenly shifting. There is also, here and elsewhere in connection with savages (recall, for example, the passage about the feast in *A Narrative of Travels*), a palpable sense of mystery and uncertainty – the need to supplement what one sees with what one imagines or guesses. Showing the mass of one's experience to be lacking, shifting the very earth beneath one's feet, producing a sense of the incomprehensible or mysterious even when directly gazed on: the "true denizen of the Amazonian forests," like an earthquake, "is unique and not to be forgotten" (*ML* I: 288).

But if this analogy between earthquakes and savages holds, we ought to find in connection with the latter an essential element of Wallace's reaction to the former: danger. Wallace rarely dwells on any threat to his physical person that savages might pose. In light of the sense of vertigo and instability to which they give rise, though, the more evident and, in some senses, fearful danger has to do with damage done to a cynosural sense of the place of Europeans on the planet, and by implication of humans in the universe. This was certainly an aspect of Darwin's response to savages: "The astonishment which I felt upon first seeing a party of Fuegians on a wild and broken shore," he wrote, "will never be forgotten by me, for the reflection at once rushed into my mind – such were our ancestors."[61] In Sigmund Freud's famous account of the "three blows" to human narcissism delivered by science, Darwinian evolutionary theory is specified as the second such blow, one that gave the lie to the notion of human uniqueness in the animal kingdom.[62] In manifold and complex ways, for Darwin himself it was his encounter with Fuegians that led to the peculiar force of that blow. But, as I show in the previous chapter, Fuegians also provided Darwin with a way to ward it off: he displaces onto them much of what he understands to be horrifying about human descent from animals.

For Wallace, too, savages at once strike such a blow to his sense of human centrality and provide the means to avert it. But he recuperates such centrality at the expense, not of savages, but of the superiority of European civilization and the contentions of evolutionary theory itself. Initially fulfilling his expectations by exceeding them, showing his imagination and prior experience to have been lacking, savages haunt him ever after. Returning throughout his life to the way they look, the way they live, what he understands to be their frequently high moral standards and admirable aesthetic production, Wallace remembers savages with an ever-increasing sense of the damage they do to his sense of the world and his place in it.

In "The Native Problem in South Africa and Elsewhere," an article published in 1906, the eighty-three-year-old Wallace looks back on his youthful travels when wishing to find evidence for the soundness of his opposition to the very forms of despotic imperialism for which, in *The Malay Archipelago*, he had once praised the Dutch. "For nearly twelve years," he recalls,

I travelled and lived mostly among uncivilised or completely savage races, and I became convinced that they all possessed good qualities, some of them in a very remarkable degree, and that in all the great characteristics of humanity they are wonderfully like ourselves. Some, indeed, among the brown Polynesians especially, are declared by numerous independent and unprejudiced observers, to be physically, mentally, and intellectually our equals, if not our superiors; and it has always seemed to me one of the disgraces of our civilisation that these fine people have not in a single case been protected from contamination by the vices and follies of our more degraded classes, and allowed to develope their own social and political organism under the advice of some of our best and wisest men and the protection of our world-wide power. That would have been indeed a worthy trophy of our civilisation. What we have actually done, and left undone, resulting in the degradation and lingering extermination of so fine a people, is one of the most pathetic of its tragedies.[63]

If, in anticipation of later forms of primitivism, such sentiments find in savage life reservoirs of much that has been lost to modern humanity, they cannot promise restoration or fulfillment. Rather, like Wallace's memory of the sinking of the *Helen* and his precious South American collections, they point to a moment of plenitude from the vantage point of its loss. For Wallace, to remember, to look back, is to see something in the process of disappearing. The inhabitants of the banks of the Uaupés will become degraded; many of those of the Malay Archipelago, extinct;

the natural respect savages feel for the rights of others in their community dwindles, in European modernity, to an unjust and tangled system of laws and their enforcement. Progress is regress. But, on the other hand, to remember the savage is also to remember what the human once was and thus what it might still become. For Wallace, recalling "savages" offers the means to recall what it is to be human as well as the promise of restoring humanity to a degenerate world.

CHAPTER 3

Charles Kingsley's recollected empire

In "Homeward Bound," the final chapter of *At Last: A Christmas in the West Indies* (1871), Charles Kingsley reflects on his deeply equivocal feelings about returning to England after a brief sojourn in the Caribbean spent chiefly on and around the island of Trinidad. Relieved to be homeward bound, Kingsley at the same time finds himself unsettled by the wish for a continuation of his voyage. South America calls him:[1]

The hunger for travel had been aroused – above all for travel westward – and would not be satisfied. Up the Oroonoco [*sic*] we longed to go: but could not. To La Guayra and Caraccas [*sic*] we longed to go: but dared not ...

But the longing to go westward was on us nevertheless. It seemed hard to turn back after getting so far along the great path of the human race; and one had to reason with oneself – Foolish soul, whither would you go? You cannot go westward for ever. If you go up the Oroonoco, you will long to go up the Meta. If you get to Sta. Fe de Bogota, you will not be content till you cross the Andes and see Cotopaxi and Chimborazo. When you look down on the Pacific, you will be craving to go to the Gallapagos [*sic*], after Darwin; and then to the Marquesas, after Herman Melville; and then to the Fijis, after Seeman; and then to Borneo, after Brooke; and then to the Archipelago, after Wallace; and then to Hindostan [*sic*], and round the world. And when you get home, the westward fever will be stronger on you than ever, and you will crave to start again.[2]

Although he has already turned toward home, Kingsley cannot help but contemplate, with a rueful backward glance over his shoulder, further travels he desires but cannot pursue. In the course of that contemplation, Trinidad takes its place not as the most southerly outpost of the British West Indies but rather as the gateway to the South American mainland and regions still further afield.[3] Envisioned in terms of its prevailing direction ("hunger for travel had been aroused – above all, for travel westward") and in connection with those who have journeyed over parts of it before (Darwin, Melville, Wallace, *et al.*), the line Kingsley draws from the Caribbean up the Orinoco River, through Venezuela, Colombia, and Ecuador, over the Andes, and out into the Pacific retraces the route of European exploration and conquest. Adopting that route as the prospective itinerary of an individual holiday tour, Kingsley does not so much trivialize the deeds of the conquistadors, naturalists, cartographers, and sailors who preceded him as aggrandize his own undertaking. Collapsing the distinction between the world historical and the modestly individual, rendering imperialism nearly indistinguishable from vacationing, he locates the impetus for both in the same restless desire for traversal of non-European space. Moreover, in shifting from the first-person plural to the second-person singular in the service of reasoning with himself, he also seems to presume his audience shares that desire: "If you go up the Oroonoco, you will long to go up the Meta. If you get to Sta. Fe de Bogota, you will not be content till you cross the Andes and see Cotopaxi and Chimborazo." Past explorers, present author, and future readers are drawn together, united by an ineluctable responsiveness to the siren song of the west. Rendered as an imperative, such responsiveness provides the title of one of Kingsley's most popular novels, exclamation point and all: *Westward Ho!* (1855).[4]

This reading of the westering urge at the end of *At Last* remains incomplete, however, to the degree that it overlooks the inscription of the yearning for occidental movement in a narrative expressly predicated upon its impossibility: "Up the Oroonoco we longed to go: but could not. To La Guayra and Caraccas we longed to go: but dared not." Kingsley imagines a route he knows he will be unable to take. And although this dynamic is most readily understood in terms of the allure of the out of reach or the forbidden, it lends itself equally well to an opposite interpretation: Kingsley may covet travel westward because it is impossible, that is, but he also, more tellingly and strangely, believes it to be impossible because he covets it. His inability to undertake the journey results directly from his desire to do so.

The workings of this perverse economy are reiterated at the end of the passage, which posits a restlessness intensified rather than assuaged by projected accomplishment. Kingsley depicts the final moment of his dreamed-of westerly circumnavigation of the globe as one in which the prospect of coming to a stop will seem further off than it did before travel began, for at that moment "the westward fever will be stronger on you than ever, and you will crave to start again." The voyage he envisions can never commence, in part because it has already been completed by others; further, were he somehow able both to begin and to finish it, the longing to which the thought of the voyage gave rise would not simply persist but grow still more powerful. Kingsley posits a wanderlust for which there can be no satisfaction – or, more precisely, in deploying the language of disease that constitutes a staple of European narratives of tropical exploration he diagnoses himself as afflicted with an incurable westward fever.[5]

At Last in its entirety structures the relation between Victorian Britain and the Americas as one of the former's insatiable longing for the latter. In one sense this longing is specific to Kingsley, for whom the tropics of the New World represent an irreclaimable ancestral land toward which he can travel but on the shores of which he can never set foot. But if *At Last* is the species, the genus to which it belongs is the one I subject to scrutiny in *Darwin and the Memory of the Human* as a whole: the Victorian construction of South America as a site of memory.[6] For Charles Darwin and Alfred Russel Wallace, the continent's status as such a site coalesces most frequently and with the most far-reaching implications around indigenous South Americans who assume the form of a living memory of the human. By contrast, the precursors Kingsley insistently recollects are sixteenth- and seventeenth-century explorers and eighteenth- and nineteenth-century natural historians, and the stakes involved in recalling them have to do less with what it is to be human than with what it was and might one day be again to be English. The desire for Britain's re-colonization of the New World, a palpable but muted presence in Darwin's and Wallace's work, informs *At Last* from beginning to end, and it is around the promises and perils of that desire that Kingsley's memory of the Americas takes shape.

Kingsley's upbringing was formative in this regard. His mother was an English Creole from Barbados, and he grew up haunted by the tale of lost family property in the Caribbean, a tale he supplemented with voracious reading in the works of exploration and settlement with which his maternal grandfather's library was stocked.[7] As he writes of the West Indies and the mainland of Central and South America on the first page of *At Last*: "From childhood I had studied their Natural History, their charts, their Romances, and alas! their Tragedies; and now, at last, I was about to compare books with facts, and judge for myself of the reported wonders of the Earthly Paradise" (*AL* 1). Anticipating the vision of endless westward movement with which the book ends, this opening formulation dooms the journey to failure from the outset. For even as Kingsley mounts an appeal to direct observation and personal experience by determining "to compare books with facts" and to "judge for [him]self," he renders such comparison and judgement superfluous because the facts in question have already been written up, precisely, as books. Flora, fauna, and geography; the displacement and extermination of native peoples; territorial battles among conflicting European powers; piracy, trade in gold and sugar and slaves, the fate of Creole plantocrats before and after the 1833 abolition of slavery in British dominions; his own family's successes and failures, voluntary and forced migrations: these and other aspects of the history of the Caribbean and Central and South America, natural and unnatural, are for Kingsley so many "charts," "Romances," and "Tragedies" – not facts merely but facts given shape by the conventions of genre.[8] As a consequence, these regions may be visited only in the way that we (or Kingsley himself) might visit, for instance, Robinson Crusoe's island.[9] Encountered in childhood in the form of romance and tragedy, not simply in texts but as textuality, Kingsley's New World possesses all the mesmerizing power of books – and all their unreality and inaccessibility as well. No wonder, then, that the shores of South America remain unvisited; no wonder that *At Last* closes with the regret attendant on unfulfilled desire.

Such exilic yearning for the New World, the details of which constitute something like Kingsley's idiosyncratic pathology, coincides with and incorporates elements of a widespread Victorian preoccupation.[10] Over the course of the nineteenth century, and often by way of that preoccupation, the Caribbean and Latin America were central to attempts at articulating what Britain's proper stance toward the rest of the globe should be. This was the case despite the relative paucity of British possessions in the region. Apart from a string of islands in the West Indies, the presence of Britons on the ground was limited to small, far-flung outposts such as the Mosquito Coast of Nicaragua, the Falklands (known to Argentines as Las Malvinas), and British Guiana, or to expatriate communities in urban centers such as Buenos Aires and Rio de Janeiro. The practical significance of such outposts, although far from negligible, pales in comparison to that of the vast military, bureaucratic, and commercial apparatus spread across North America, Africa, Asia, and Oceana, an ever-expanding formal empire that would, by the early twentieth century, extend over more than a quarter of the land surface of the globe. The Caribbean and Latin America, however, carried an imaginative weight out of all proportion to their relative unimportance as components of that empire. To take merely two examples: efforts to persuade Brazil to outlaw slavery (legal there until 1888) provided enduring opportunities for Britons to portray themselves as occupying an unassailable moral high ground on that issue, while the case of Governor Eyre's suppression of the Jamaican Uprising in 1865, even as it sharply divided opinions at home on questions of race and colonial rule, served as the occasion for a very public exercise in national self-fashioning.[11]

Kingsley's *At Last* constitutes a signal instance of such attempts at self-fashioning by way of the Americas. It belongs to a group of travel narratives, historical novels, and popular as well as academic histories produced from mid-century on that attend not to the persistence of the slave trade in the present or to those territories in the region formally occupied by the British but to the wider historical and geographical contours of Britain's engagement with the New World. One such text, Anthony Trollope's *The West Indies and the Spanish Main* (1859), in other ways quite different from *At Last*, tellingly resembles Kingsley's book in the scope promised by its title, which, in pairing "the West Indies" with the anachronistic designation "the Spanish Main," conjoins two regions that by the 1850s might well have been thought of as entirely distinct.[12] The travel writing, fiction, and history to which I refer above effects the same slippage by concerning itself with an era in which the Caribbean islands and the Central and South American mainland were united by virtue of being a vast arena of conflict, neither British nor "Latin" but still in the process of being fought over. Invoking a New World in the Americas apparently full of opportunity for Britons, these texts allow the figurative incorporation of the region into the British Empire and, crucially, the solidification of Britain's claim to being an imperial power from its beginnings and in its essence.

The recourse to such temporal sleight of hand in the context of nineteenth-century Latin America was rendered possible because of its status, from the point of view of the British, as a literal anachronism. Together with a handful of other non-European parts of the world, Latin American nations often figured as a disruption of British imperialist discourse during the Victorian era.[13] They disabled the progressive narrative of empire by appearing to be anachronistic, not simply vis-à-vis Europe, but in relation to the conventions of that narrative itself. As the remnants of former (Spanish and Portuguese) empires, they stood as a monument to colossal imperial failure, a potential *memento mori* for imperial aspirations. As parts of the world in which British imperialism rarely took the spectacular form of military occupation and settlement but instead proceeded along less visible lines such as banking, government loans, mining concessions, and railways, they remained difficult to assimilate into the heroic story of the Anglicization, Christianization, and civilization of savage lands. As territories that had been "discovered" in the fifteenth century, they transformed tales of supposed first contact into fables of belatedness.[14] Thus it is that Victorian writing about Latin America so often falls into modes of lamentation and prognostication.[15] Itself deeply inflected by a tone of lament, Kingsley's *At Last* nonetheless embraces Latin America as an anachronistic space in the service of recuperating it for the narrative of British imperial expansion.

Of the group of texts that pursue a similar undertaking, by far the most influential, as well as the one in which the pattern to be found in *At Last* is set forth most starkly, is *The Expansion of England* (1883), Sir John R. Seeley's effort to explain how the inhabitants of a small island in a corner of the North Atlantic came to exercise sovereignty over millions of subjects scattered across several continents. Seeley argues that Britain's historical engagement with the Americas gave rise to its modern imperial identity. This argument, in turn, provides him with evidence for the much more far-reaching assertion that Britain and its colonies should be understood as one immense expanded England or "Greater Britain."[16] According to *The Expansion of England*, that is, Britain is the New World.

Perhaps because he turns that formulation around to contend that the New World is Britain, Kingsley is at once more invested in and more troubled by the project of redefining "home" to include the Americas. Although fundamentally in sympathy with Seeley's attempt to make of the Caribbean and South America a once and future empire, Kingsley's own efforts at doing so depend so entirely on retrospection and longing that, as in the hallucinatory evocation of westward travel that comes at the close of *At Last*, he can only offer up such a prospect in a form that ensures the impossibility of its realization. Neil L. Whitehead identifies a closely related vision in the accounts of twentieth-century travelers to South America: "Each journey recapitulates the defining experiences of tropical travel – the discomfort of the traveller, the disingenuousness of the natives, and ultimately a disgust at the weakness of the self – which can then be expressed as a nostalgia, but for something that never was."[17] Anticipating this structure of feeling, Kingsley at the same time demonstrates its potential productivity. In his hands, nostalgia for a past that never was takes the form of longing for a history he never experienced, a history that, by allowing Kingsley's voyage to be figured as a repetition and reflection of the past, dwarfs that voyage while simultaneously and paradoxically guaranteeing its gravity.

In this way Kingsley's unassuming travel narrative casts into high relief the opportunities Latin America afforded – as well as the difficulties it posed for – Victorian imperial imaginings. Decidedly marginal to any project of formal empire in the nineteenth century, Latin America in Kingsley's account, in part because of its marginality, enables the elaboration of an imperial identity for Britain and for certain Britons in ways that the extant formal empire does not. At the same time, it cannot help but foreground not only the intensely artificial nature of that identity, constructed or posited rather than given, but also its tendency to alienate Britain from itself, to install at its heart a backward-looking desire for which there can be no fulfillment.[18]

In their travel writing and works of evolutionary theory, Darwin and Wallace constitute the human as such around a similar desire. Like Kingsley's national subject, the subject of evolution as they elaborate it depends on a past that must be recalled but that cannot be returned to. Kingsley lays bare the potential complicity of such a mnemonics with the work of nation and empire. At the same time, he foregrounds the insuperable impediment it poses to that work. Recalling the past in the service of redressing failings in the present, he displaces all that is vital and powerful to an anterior time that may be invoked but cannot be relived. Which is just to say that, if Kingsley assimilates the savage past into his retrospective fantasy of imperial expansion, the price of doing so amounts to the permanent deferral given memorable expression in the passage from *At Last* with which I began this chapter.

I

Recent accounts of nineteenth-century imperialist discourse compellingly argue for the paradigmatic status of what Johannes Fabian calls the "denial of coevalness," a practice of thinking and writing that represents Europeans as occupying a historical "present" from which non-European peoples are excluded insofar as they are understood to be out of date, behind the times.[19] As I discuss in the Introduction, Anne McClintock names one of the textual manifestations of that practice "anachronistic space," a trope according to which "colonized people ... do not inhabit history proper but exist in a permanently anterior time within the geographic space of the modern empire as anachronistic humans, atavistic, irrational, bereft of human agency – the living embodiment of the archaic 'primitive.'"[20] Victorian fiction, travel writing, and ethnography regularly feature more or less sophisticated variants on anachronistic space. Its *locus classicus* is perhaps Marlow's description of Africa in Conrad's 1899 novella, *Heart of Darkness*: "We were wanderers on a prehistoric earth, on an earth that wore the aspect of an unknown planet ... We could not understand, because we were too far and could not remember, because we were travelling in the night of first ages, of those ages that are gone, leaving hardly a sign – and no memories."[21] In Marlow's account, the steamer he pilots slowly up the Congo functions as a time machine, returning him to a past so distant that it has left "hardly a sign – and no memories." The utter unfamiliarity of this "earth that wore the aspect of an unknown planet" insists on temporal rather than merely spatial dislocation: unrecognizable as belonging to the same world as that modernity for which Marlow serves as reluctant emissary, the African interior can only register as the space of a prehistoric "before," preceding and so standing outside of the flow of time.

Latin America, too, often figures in British accounts as such an anachronistic space. It is no accident that the group of "sordid buccaneers" Marlow encounters in central Africa call themselves the "Eldorado Exploring Expedition" – or that the novel that pushes this trope to its absurd limits, Arthur Conan Doyle's *The Lost World* (1912), takes the South American rainforest as the natural place to look if one is in search of still-extant dinosaurs and proto-humans.[22] But if Doyle's novel and similar texts represent Latin America as anachronistic in McClintock's sense, out of step with Europe and so, like many other non-European areas, subject to "discovery" and conquest in the name of modernization, they also trade in another kind of anachronism. At a time when most of those other areas were steadily being brought under direct control by European powers, a trend that reached its peak with the frenzied Scramble for Africa of the last two decades of the nineteenth century and the first two of the twentieth, Latin America was largely spared formal seizure. As P. J. Cain and A. G. Hopkins point out in *British Imperialism*, the reasons for this were both pragmatic and programmatic: "There was, of course, no formal partition of South America. Considerations of cost, logistics and diplomacy were always on hand to restrain the major powers at moments of crisis. Far more important, however, was the fact that official political intervention was rarely demanded by economic interests; nor was it seen by the Foreign Office to be appropriate."[23]

The determining influence given to economic interests in this account signals the direction of Cain and Hopkins's re-theorization of British imperial expansion as driven less by ideological and commercial designs for which seizure of territory was necessary than a "gentlemanly capitalism" subtended by a demand for income compatible with informal as well as formal relations of dominance – a re-theorization for which, they claim, "Britain's relations with South America in the nineteenth century [provide] the crucial regional test."[24] Other historians seeking to describe Victorian Britain's involvement with the then-new republics of Central and South America have had recourse to similar formulations: "informal empire," first suggested by Ronald Robinson and John Gallagher, or "business imperialism," coined by D. C. M. Platt. Both qualifications – "informal" and "business" – not only mark a difference from some other, presumably more formal politico-military form of dominance but also suggest that British practice in Latin America serves as a precursor to what has been considered a decidedly later phase of imperialism: neo-colonialism.[25] Put forward at the end of the twentieth century as a revision of heretofore standard accounts of the workings of empire, Cain and Hopkins's emphasis on economics rather than usurpation of land and sovereignty, and thus on what might be thought of as the anticipatory shape of Victorian imperialism in Latin America, was registered a century before in texts ranging from Doyle's popular romance to Conrad's proto-modernist tour de force, *Nostromo* (1904).[26]

In economic and political as well as discursive terms, then, Latin America interrupts the smooth forward motion of the story of the Empire. It resists in particular the teleology of imperial narrative, appearing in relation to the assumptions of that narrative at once behind the times – despite centuries of European influence, still prehistoric or savage – and ahead of them – both as the remains of former Spanish and Portuguese empires that threatened to render the nineteenth-century British Empire archaic, already outdated, and as an area in which the British practiced neocolonialism at the height of the colonial period. From about 1850 on, a series of texts appeared that, in diverse and sometimes contradictory ways, confront the apparent temporal anomalies of the Americas and attempt to parlay them into support for empire and for a conception of Britain as essentially imperial.[27] Of these various and variously successful works, Seeley's *The Expansion of England* was arguably the most important.

First delivered in 1881 and 1882 as a series of lectures at Cambridge, where Seeley held the Chair of Modern History, *The Expansion of England* was published in book form in 1883. It had an impact both immediate and long lasting. Widely read at the time of its publication, it remained uninterruptedly in print into the second half of the twentieth century.[28] Best known now as the location of the infamous pronouncement that "We [Britons] seem, as it were, to have conquered and peopled half the world in a fit of absence of mind," *The Expansion of England* actually concerns itself less with acquisitive absent-mindedness than with driving home "the simple obvious fact of the extension of the English name into other countries of the globe," a "fact" that Seeley insists amounts to "the foundation of Greater Britain."[29] Taking as his point of departure the "fantastic … notions of abandoning the colonies or abandoning India, which are so freely broached among us," Seeley reviews the history of Britain's overseas involvement in such a way as to portray the granting of independence to colonial possessions as a dismemberment of the body of the nation-state (*EE* 241). While he sharply distinguishes the settler colonies – "our colonial empire" – from India, and devotes the entirety of the second half of his book to the distinctive problems posed by considering India as part of an expanded England, Seeley concludes that neither those colonies nor India itself should be seen as separate or separable from Britain (*EE* 14, 241).[30] Although far from universally embraced, this argument for retaining imperial possessions was widely popular, particularly with the leading imperialists of the final years of the nineteenth century. By the 1890s it was to be adopted by those men Deborah Wormell calls the "representatives of *fin-de-siècle* imperialism," including "Joseph Chamberlain, Cecil Rhodes, Lord Rosebery and W. T. Stead."[31]

If that list of names adumbrates the role of *The Expansion of England* in providing a rationale for the New Imperialism, among the book's most notable features is the astonishing place it assigns to a decidedly "old imperialism" in the Americas, a region that by Seeley's lights must be seen as the crucial factor in the evolution of the expanding England of his title. History provides access to the truth of Britain's identity as an imperial power to the extent that it promises to locate the origins of that identity, origins that are to be found long before the nineteenth century and far from Europe. Leaving to one side the story of Victorian imperial hegemony, a story set largely in Africa and Asia, Seeley returns to an earlier period in which the Americas were of paramount importance. The reign of Elizabeth I marks the beginning of that period, for "this was the time," he writes, "when the New World began to exert its influence, and thus the most obvious facts suggest that England owes its modern character and its peculiar greatness from the outset to the New World" (*EE* 71). In this remarkable account, neither William the Conqueror's victory at Hastings in 1066, nor the Glorious Revolution of 1688, nor any other event that took place on British soil is to be seen as marking the birth of modern Britain. The England of the second half of the nineteenth century, the center of an immense Greater Britain, must look to Christopher Columbus for an unlikely founding father, for it owes its existence to the pan-European territorial struggles set off by his arrival in the New World on October 12, 1492 (*EE* 99).

Such a rewriting of England as historically and so essentially imperial exploits the potential of analepsis to insert temporally (and in this case also geographically) distant events into the narrative of present expansion.

In Gérard Genette's narratological taxonomy, analepsis refers to "any evocation after the fact of an event that took place earlier than the point in the story where we are at any given moment."[32] For Genette, analepsis and its opposite, prolepsis, constitute the two possible forms of anachrony or disjunction between the order of events in the "story" on which a narrative is based or alludes to and the order of events in the "plot," which is to say the story as it is narrated. By means of analepsis, Seeley exposes the contemporary realities of Britain's presence in the Americas as lacking even as he enriches that presence by an appeal to a more lustrous past. At its most sweeping, *The Expansion of England* turns history to account by locating in Britain's former greatness the appropriate rebuttal to proposals for abandoning its current possessions. That former greatness demonstrates that empire *is* England, that the two cannot be separated or even logically distinguished from one another.[33]

What might otherwise appear to be a fairly straightforward invention of tradition takes a startling turn, however, when Seeley makes explicit the full force of that seemingly obvious identification. Far from merely having been shaped by the New World, England should properly be considered a part of it: "In fact, as an island, England is distinctly nearer for practical purposes to the New World [than is the rest of Europe], and almost belongs to it, or at least has the choice of belonging at her pleasure to the New World or to the Old" (*EE* 77). Despite the tremendous distances that separate them from London, that is, Jamaica, Trinidad, British Honduras, British Guiana, and the Falklands should be seen as so many more recently settled versions of Cornwall or Yorkshire – largely because Seeley's "England" comes into being in its battles with Spain in the Caribbean and along the South American coastline and in its victory over the Spanish Armada.

Moreover, as a consequence of what the title of one chapter of Seeley's book names the "Effect of the New World on the Old" (*EE* 64), Britain itself has been wrenched free of its determinate Europeanness and remade as one more island in the Atlantic. Begun in the sixteenth century, the Americanization of Britain is brought up to date and completed with the successful Latin American independence struggles of the early nineteenth century. "The result of these mighty revolutions," Seeley avers, "is that the Western states of Europe, with the exception of England, have been in the main severed again from the New World" (*EE* 46). In 1824, British Foreign Secretary George Canning had asserted about the new nations of the Americas: "Spanish America is free and if we do not mismanage our affairs sadly, she is English."[34] By 1883, Seeley can suggest, instead, that England is Latin American.

Confronting Latin America as an anomaly, as the site of backwardness and atavism as well as of anticipation and futurity, Seeley has recourse to his own tortuous narrative temporalities. Interested in defending a present-day empire in which the region plays an ambiguous and less-than-spectacular role, he gives himself over to the past, to a period when Britain was very visibly engaged in subduing it. More than simply recovering Britain's imperial interest in Latin America, he elevates it to a position of unwonted prominence by tracing the British Empire – which is, for him, Britain as such – back to the moment of its most spectacular engagement there. In overcoming the resistance it posed to the narrative of British imperialism, that is, Seeley installs Latin America at the heart of the empire and of Britain's imperial identity.

2

Nearly thirty years before the publication of *The Expansion of England*, Kingsley himself had anticipated many of its strategies in a work I have already mentioned, a historical novel set in the same period Seeley finds so consequential: *Westward Ho!*, which bears the lengthy explanatory subtitle *or The Voyages and Adventures of Sir Amyas Leigh of Burrough, in the County of Devon, in the Reign of Her Most Glorious Majesty Queen Elizabeth*.[35] The ideological import of *Westward Ho!* is decidedly overdetermined. Relentlessly anti-Catholic from start to finish, the text most clearly functions as one volley in Kingsley's ongoing battle against John Henry Newman and the Oxford Movement.[36] In addition, as a work celebrating the virtues of martial Englishmen engaged in just warfare against the territorial designs of a cruel and un-Protestant enemy, the novel aspires to speak directly to the sufferings undergone by British troops in the Crimean War – sufferings that, at the time of the novel's composition and publication, were the focus of intense public scrutiny. Read in light of Seeley's *The Expansion of England*, however, the swashbuckling and treasure hunting of which much of the plot is comprised, otherwise the most hackneyed features of boys' adventure fiction, take on a particular resonance. Chiefly concerned with the exploits of Drake, Ralegh, Grenville, and other English "sea dogs," *Westward Ho!* constitutes a particularly dramatic instance of the analeptic enlistment of the "pastness" of Britain's engagement with Latin America in the service of Victorian imperialism. Dedicated to two servants of empire – Sir James Brooke, Rajah of Sarawak (the same Brooke who makes up one of the list of those who have gone west before Kingsley in the final chapter of

At Last), and the Reverend George Augustus Selwyn, Bishop of New Zealand – and replete with labored comparisons between the exploits of actual and fictional Elizabethans and their nineteenth-century avatars, Kingsley's novel foregrounds its own ambition to turn the past and the Americas to account in rationalizing Greater Britain. As David Armitage puts it in *The Ideological Origins of the British Empire*, unwittingly channeling Seeley and Kingsley: "The originating agents of empire were the Elizabethan sea-dogs, Gloriana's sailor-heroes who had circumnavigated the globe, singed the King of Spain's beard, swept the oceans of pirates and Catholics, and thereby opened up the sea-routes across which English migrants would travel, and English trade would flow, until Britannia majestically ruled the waves."[37]

At Last, by contrast, at first appears to furnish a more familiar version of the mid- to late-nineteenth-century West Indies and South American mainland according to the well-established conventions of Victorian travel writing. For it deploys that system of reference that Mary Louise Pratt has termed "anti-conquest": an apparently neutral and innocent set of natural-historical descriptions of foreign lands and peoples that explicitly distinguishes itself from earlier modes of expansionist writing while implicitly furthering European territorial and economic dominance.[38] Much of *At Last* exemplifies the traits of anti-conquest narrative. Himself an accomplished naturalist, Kingsley devotes a great deal of effort to the task of detailing the natural productions of the West Indies. More than half of the illustrations in the book, twenty-three out of forty-one, represent plants. That many of those plants happen to be food crops – guava, yam, sweet potato, breadfruit, and so on – confirms the close link Pratt suggests between scrutiny of the natural world and implicit attentiveness to the potential for settlement and concomitant agricultural production.[39] But the book's investment in negotiating a complex set of temporalities makes it difficult to read as a straightforward rationale for expansion, even an implicit one. Like Seeley, and like his earlier self in *Westward Ho!*, Kingsley in *At Last* puts history to work in the construction of an idea of the Americas as part of Greater Britain. This appeal to the historical past, however, draws upon and becomes conflated with an appeal to memory and his own personal past. The resulting text, while infusing Seeley's potentially dry, abstract notion of an expanded England with powerful libidinal energies, reveals a deeply ambivalent stance toward the project of redefining "home" so as to encompass the New World.

"At last we, too, were crossing the Atlantic": so reads the first sentence of *At Last: A Christmas in the West Indies* (*AL* 1). "At last": ambition realized;

something patiently anticipated finally come to pass; long-frustrated aspirations accomplished in the end. In choosing these two words as the first words of the text, and of the text's title before them, Kingsley places his entire narrative under the sign of culmination even as he also invokes the heretofore-unsatisfied yearnings that constitute culmination's necessary prelude. Travel narratives, to be sure, nearly all trade in expectation and retrospection: one goes to the other place because it is imagined to be the locus of the desired; one writes of the trip in retrospect, having been where one wished (if not always having found what one sought). But Kingsley's text seizes upon such generic conventions with particular ferocity. The Caribbean and the "Spanish Main," he relates, is a place of romance and tragedy like no other. As the first chapter unfolds, the journey the text documents takes on the status of that which will give the years that went before their value and meaning. Landfall in the New World, far from a moment of mere holiday fun or the opportunity to satisfy idle exoticist curiosity, amounts to nothing less than "the dream of forty years translated into fact" (*AL* 18).

Those "forty years" of Kingsley's life demarcate a long and varied career that included success as a novelist, poet, clergyman, naturalist, Christian socialist, and above all as the most vociferous enemy of the Oxford Movement and proponent of the so-called muscular Christianity that advocated a robust, "manly," Protestant embrace of the material world. The first interpretive crux in *At Last*, then, has to do with discerning how it might be that a three months' steamer journey from England to the Caribbean and back again could stand for Kingsley as the thing about which he writes "at last." In what sense can such a trip be understood as the fitting conclusion to the life of someone biographer Susan Chitty calls "undoubtedly one of the most prominent men in England in his century," a man who "achieved distinction, or, in some cases, notoriety, in many fields"?[40] An initial solution to this problem lies in the particular hold the tropics have on Kingsley's imagination. In spending his Christmas in the West Indies, he comes into contact with ground consecrated by ancestral presences, walks through scenes of what he repeatedly refers to as an "Earthly Paradise" first encountered in books and family stories in childhood but dreamed of (and read and written about) ever since (*AL* 1, 25, 27, etc.).

The text's first sentence indicates this much as well: "At last we, too, were crossing the Atlantic." "[W]e, too": the trip documented in *At Last* serves as a culmination not merely because it has been long desired and delayed but also because it repeats the journeys of others.

Quite contrary to those voyages of discovery in which the drive to be first is paramount – as, for instance, the earnest and bitterly contested search for the source of the Nile carried out by Richard Burton and John Hanning Speke, among others – Kingsley's voyage is expressly undertaken and represented as repetition. As Simon Gikandi writes, "Kingsley's narrative eschews the authority of original experience and revels in its secondariness."[41] Kingsley takes pleasure, we might say, in not so boldly going where many men have gone before.

Everything encountered is, if not actually new, at least new to him, and so occasionally the language of discovery slips in: December 17, 1869, for instance, is referred to as the day on which he "first sighted the New World" (*AL* 14). Such appearances in the text of the rhetoric of coming first, however, do not signal any genuine primacy so much as an intense identification with the exploits of precursors. Exhilaration inheres in repetition. Far from a compromise formation, some thinly veiled attempt to make a virtue of the hard necessity of belatedness, such a replacement of the desire to be and see first with that of being second and seeing again constitutes Kingsley's preferred mode of encounter.[42] To explain his delight at crossing the Atlantic he dwells upon the fact that he is "on the old route of Westward Ho" or "on the track of the old sea-heroes" (*AL* 2, 5). As he leaves Port of Spain, an otherwise uninteresting bit of scenery becomes worthy of comment when he recognizes it as the "very spot [where] Raleigh and his men sailed in to conquer Trinidad" (*AL* 148). Crossing an area of open water near St. Thomas, he alerts readers that "[t]his channel has borne the name of Drake, I presume, ever since the year 1575" (*AL* 392). At book's end, as we have seen, he sketches out the travel he desires but cannot undertake in relation to Darwin, Wallace, and other men who have already completed it. And so on: similar pronouncements punctuate the text over the entirety of its 400 pages. Belatedness, usually the worst disaster that can befall an explorer, is for Kingsley something to be embraced.

Why this should be so has to do with the peculiar convergence in *At Last* of memory and homosociality. Like so many of his compatriots, Kingsley encounters the Americas as a *lieu de mémoire* or "site of memory." The recollections to which that site provides access, however, are for him neither corporeal – the sort of involuntary, bodily memory of savagery to be found in the work of W. H. Hudson, for instance – nor entirely those of others – as they are for Darwin in his relation to Alexander von Humboldt, or Wallace in his relation to Darwin. Of course, what Kingsley recalls about Trinidad, St. Thomas, or the South

American mainland does not derive from his own first-hand experience; despite taking shape around the fact of Kingsley's belatedness, *At Last* nonetheless does literally document his first sight of these places (or, in the case of South America, his failure to secure a sight of this place). His recollections are of his encounter with others' recollections, and in this sense they both do and do not belong to him. Further, if the source of those recollections is demonstrably maternal, since Kingsley's mother serves as his most immediate point of contact with the lost Eden in which he finds himself, as they play themselves out in his text they are exclusively, insistently gendered male. Not a member of an extended brotherhood so much as the latest representative of a male line of succession stretching back into the distant past, Kingsley secures his belonging by positioning himself to recall, in his own person, what his precursors first saw – and what only they could properly have remembered.

In his work on the cartographer Robert Schomburgk and other nineteenth-century travelers to South America, D. Graham Burnett names a relation to predecessors similar to Kingsley's with the word "metalepsis": "invocations of history that are deployed in order to authorize even as they are stripped of their authority and content."[43] In *Masters of All They Surveyed*, Burnett contends that these "later travelers cycle back over their predecessors by editing their texts and redrawing their routes on fresh maps, all the while borrowing orienting points, geographical authority, and ennobling historical context from them."[44] Draining originary moments of their hold on the present by redescribing them as primitive or elementary stages in a narrative of development, metalepsis allows otherwise belated travelers to overcome their predecessors even as they profit from them. Kingsleyan analepsis, like Schomburgk's metalepsis, marshals the past to authorize the present. But in Kingsley's case such authorization also demands that the past displace him, rendering him superfluous or supernumerary. His deployment of analepsis ensures that the sense of coming "after" entirely defines his construction of himself as traveler.

The same sense may also be said to characterize the land to which he travels. If it cannot precisely be described as belated, on Kingsley's view that land certainly seems to live something like a posthumous existence. At times his feeling of the pastness of real life and vitality is so powerful that it is as if the Caribbean islands were peopled not so much with living inhabitants, the "Negros [*sic*], Coolies, Chinese, French, [and] Spaniards" he notes upon arrival in Trinidad, as with the dead – and the English dead in particular (*AL* 66). He repeatedly invokes those of his countrymen who have come before and left their remains behind, "ghosts of gallant soldiers and sailors" who, though now long gone, have no choice but to haunt the living Englishman literally and figuratively following in their wake (*AL* 43). Especially when attempting to ward off the possibility that Britain will countenance various drives for independence on the part of its possessions, Kingsley resorts to a necromancy of national memory, summoning these ghosts so that he can speak for them and as them to repudiate calls for relinquishing territory or sovereignty: "But was it for this," he asks, or has those who have come before him ask, "that these islands were taken and retaken till every gully held the skeleton of an Englishman?" (*AL* 43).

The allusion to the great nineteenth-century poem of memory, William Wordsworth's *The Prelude* (1850), is instructive. In some of his most famous lines, Wordsworth asks:

> Was it for this
> That one, the fairest of all rivers, loved
> To blend his murmurs with my nurse's song,
> And, from his alder shades and rocky falls,
> And from his fords and shallows, sent a voice
> That flowed along my dreams?[45]

The enigmatic "this" refers to the poetic crisis outlined in the immediately preceding section of Book First: the speaker's inability to find a suitable theme for the "arduous work" of writing he wishes to undertake.[46] The entirety of the poem turns on these six lines, for they mark the moment at which the speaker first looks to his past in search of a solution to present troubles, establishing memory as the ground for all that follows. The memory in question is of the speaker's own childhood and youth, and the solution arrived at "the discipline / And consummation of a Poet's mind" referred to in the poem's final book.[47] Thus, although James Chandler illuminates the larger implications of such discipline by demonstrating its status as the "psychological manifestation of a national character and a native tradition," it is in the first and last instance personal.[48] Only as they originate in Wordsworth's early days, and only as the poet recalls them at a time of crisis, do the memories of which *The Prelude* is made up establish the Burkean Englishness that returns the poet to himself and, in so doing, enables him to commence the great work he falters before in Book First.

Kingsley's recollections, however, can only be considered those of his own childhood at one remove, and the antecedent of his "this" is not a poetic but a political crisis: Britain's apparent willingness to give up the territory for which so many of the predecessors whose memories he cherishes and whose journeys he repeats died. Inverting the mnemonic logic of *The Prelude* by constituting himself as the bearer of second-hand national and imperial memories, Kingsley also reforms the story it tells. Wordsworth's poem, although routed through the nation, commences with and returns to the poet; Kingsley's travel narrative, routed through Kingsley himself, begins and ends with the imperial nation-state.

Finally, then, *At Last* may be understood as a text that raises the specter of not one but two pasts: an individual past of reading about, writing about, and longing for the tropics and an imperial past of the exploits of English explorers and buccaneers in the New World. These two pasts, these two specters, become in the course of the narrative one and the same. Conflating the narrative analepsis or leap backward to his own boyhood with a similar leap back to a period troped as the boyhood of imperial Britain, representing his holiday journey to the West Indies as a redemptive voyage through time as well as space, Kingsley makes possible a reading of the "at last" of his title as a personal expression of longing fulfilled that is, at the same time, a national-corporate expression of destiny accomplished.[49]

3

Not all the pleasures and paradoxes of the New World, however, have to do with Kingsley's sense of himself as living out a boyhood dream or encountering a region made precious by the sacrifices of so many English heroes. At least as prominent a feature of the text is that natural-historical observation mentioned above in connection with Pratt's notion of anti-conquest writing. Consider, in this regard, the moment of arrival on St. Thomas, the site of Kingsley's first view of the Americas, when he refuses an invitation to tour the island's principal settlement. "To town?" he asks, inflecting his prose with the same disdain he must have visited upon the hapless islander who extended such an unfortunate invitation. "Not we, who came to see Nature" (*AL* 17). He devotes his initial steps on the hallowed ground of the West Indies to the exploration of their natural productions – productions which, when encountered, shock and delight with their "wonderful wealth of life" (*AL* 18). Everything, he declares again and again, is "utterly new and strange," cause for "astonishment" (*AL* 95, 18). The gullies of these islands hold the skeletons of Englishmen clamoring for the visitor's attention, but they also hold what appears to be an equally compelling array of plants and animals: "every [natural] form was demanding, as it were, to be looked at" (*AL* 18).

Cataloguing the delights of tropical nature, however, seemingly so far removed from the return to the land of childhood or the rehearsal of English imperial history that characterizes much of the rest of the text, shares the same structure of belatedness and retrospection – and reaches the same equivocal culmination, which stands as a marker for its own impossibility. As in his attentiveness to historically significant landmarks, so in his admiration of the natural world on view: Kingsley never ceases to embrace his role as follower. He notes of his reasons for setting out on the voyage in the first place that he "longed to behold, alive and growing, fruits and plants which [he] had heard so often named, and seen so often figured" (*AL* 372). As Wallace had discovered during his time in South America and Southeast Asia, the lot of the Victoria natural historian required constantly confronting the dissonance of encountering "for the first time" exotic organisms that had in fact been encountered many times before – by others but also by the natural historian himself in travel narratives, the plates of natural history books, museum collections, hothouses, and menageries. When he finds himself in the "primeval forest," for instance, Kingsley takes pains to point out that he "look[s] upon that which [his] teachers and masters, Humboldt, Spix, Martius, Schomburgk, Waterton, Bates, Wallace, Gosse, and the rest, had looked already, with far wiser eyes than mine" (*AL* 157).[50] But whereas for Wallace the fact of nearly always having to see second produced disappointment, for Kingsley it promises instead a kind of privileged second-sight that heightens rather than diminishes the wonders on offer. Wallace deplores having been misled by the hyperbole of earlier explorers; Kingsley proclaims that "[w]hat I had heard and read is not exaggerated" (*AL* 303).

It is Wallace himself, along with the other travelers and natural historians Kingsley lists, who sets the pattern for his fascination with flora and fauna so different from that of England. Further, Wallace and the rest are in part responsible for the fact that, like so much else encountered in the New World, those flora and fauna are known in advance – not discovered but recognized, not collected so much as recollected, as when exotic sea shells picked up on the beach figure as both remarkable surprises and, predictably but also jarringly, "old friends in the cabinets at home" (*AL* 20). Edward Said has observed that this pattern of knowing before going is typical of the eighteenth- and nineteenth-century European traveler, whose task it is not so much to discover the new as to confirm what has already been established.[51] For Kingsley, however, the familiarity of what he encounters functions more specifically than merely as such a confirmation. Rather, as his description of the shells he finds suggests – and here is

Figure 1 "The Botanic Gardens, Port of Spain," frontispiece, Charles Kingsley, *At Last: A Christmas in the West Indies* (London: Macmillan, 1871).

the point of closest convergence between Kingsley and Seeley – the New World and its denizens are familiar because they are in some ways understood to be home.

Exemplary in this regard is the frontispiece to *At Last*, an illustration depicting, according to the caption, "The Botanic Gardens, Port of Spain" (Figure 1). The scene is an empty, almost a desolate one. No human figures appear, although the bare outlines of buildings visible in the background, overshadowed by foliage in the center and near the left-hand side of the picture, suggest the possibility of their presence. Trees dominate the visual field, stretching from left to right over the entirety of the middle ground. Elsewhere Kingsley specifies that these gardens contain "trees from every quarter of the globe," and thus presumably examples of many different species, genera, and families (*AL* 94). The frontispiece, though, represents the gardens as given over almost entirely to palms. Since they are emblematic of the tropical nature Kingsley traveled so far to experience first-hand, trees, and palm trees in particular, might seem to serve as a fitting introduction to and summation of the journey documented in *At Last*.[52] Given the wealth of other possibilities, however, the choice of "The Botanic Gardens, Port of Spain" for the frontispiece is a curious one. More obvious and apparently more appropriate would have been an illustration such as the one captioned "Waiting for the Races" (Figure 2), since Kingsley makes constant reference to the racial mélange to be found on Trinidad. Or again, given Kingsley's frequently expressed

Figure 2 "Waiting for the Races," Charles Kingsley, *At Last: A Christmas in the West Indies* (London: Macmillan, 1871).

wish for something like the life of a castaway, we might expect an illustration along the lines of "A Tropic Beach" (Figure 3), a representation of the un-English beauty he has kept in his mind's eye all his life but only now comes into contact with.

In lieu of the carefully delineated costumes and postures of non-European Trinidadian races or riotous vegetation cascading down the hills that surround an isolated cove populated by a single human figure, however, Kingsley provides as readers' first glimpse and perhaps enduring impression of the New World a group of trees – trees, moreover, ranged behind a fence. We seem to be in the tropics (the palm trees see to that), but the tropics collected, managed, displayed, walled off. Palms behind a fence: a world accessible to the eye but forbidden to the rest of the body, a land of the familiar exotic, recognized from representations in books and apparently just as untouchable as those representations – indeed, reproduced as yet another such representation. Considered in connection with the familiarity of the flora in front of which it stands, however, the fence in this illustration conveys more than mere inaccessibility. It, too, is recognizable from home. Were it not that the open sky occupies the space that should be filled by glass panels, that is, we might take this fence and its walling-off of tropical vegetation as license to imagine ourselves gazing at a site on the outskirts of a British rather than a West Indian capital city: not "The Botanic Gardens, Port of Spain" but the Royal Botanic Gardens, Kew.

Figure 3 "A Tropic Beach," Charles Kingsley, *At Last: A Christmas in the West Indies* (London: Macmillan, 1871).

At last, the relief expressed in the title is legible as the relief of home-coming. The transports of joy upon viewing (familiar) tropical nature for the first time, the moments of English military and maritime history conjured up by way of their association with specific locations, the constant references to the throng of English dead, the crossing of the Atlantic as boyhood dream realized, the deliberate recollection and repetition of the exploits of precursors, the summation of the voyage and the narrative of that voyage by means of palm trees on display: all converge in producing the New World as home, as indissolubly part of Kingsley and so of Britain itself.

At several points Kingsley goes so far as to literalize that figurative convergence by imagining the possibility of becoming a colonist, of making the islands his actual home and so effecting his own modest expansion of England. Of a visit to a collection of farms in Montserrat, a region of central Trinidad, he writes: "One clearing we reached – were I five-and-twenty, I should like to make just such another next to it" (*AL* 241). But here, at just the moment when the metaphorical seems on the verge of becoming actualized, the limits of this particular construction of Greater Britain reveal themselves. "*Were* I five-and-twenty": the culmination reached in *At Last* remains firmly in the realm of the subjunctive. For to make the New World home would not simply be to extend England's reach but also to abandon (little) England and, more importantly, to abandon a powerful idea of that England as home. Elsewhere we read of clearings such as the one Kingsley would make were he five-and-twenty that they are "exquisite little land-locked southern coves," "places to live and die in" (*AL* 127). One of these coves houses "a gallant red-bearded Scotsman, with a head and a heart; a handsome Creole wife, and lovely brownish children, with no more clothes on than they could help" (*AL* 129). Identifying the Scotsman as "[a]n old sailor and much-wandering Ulysses" (129), Kingsley alludes not only to *The Odyssey* but also to Alfred Tennyson's "Ulysses" and "The Lotos-Eaters," bodying forth the satisfactions of a life on Trinidad in such a way as to render those satisfactions impossible in that they would amount to the same languorous abandonment of duty and perpetual postponement of actual homecoming to which, in the latter poem, Odysseus and his men almost succumb.[53]

Lingering in the islands, perpetually contemplated, is perpetually postponed, and for this reason: so central is the past in constituting them as home that it rewrites that home as habitable only by ghosts – a land not for Kingsley and his generation but for those who came before, who live on as memories, and, just possibly, for those who are yet to come. Viewing every new scene as a palimpsest of past and present, a space haunted by history and longing, Kingsley imagines the Americas as a place that could have been and may yet become home, but is not such now. Thus the text closes upon Kingsley's return to Britain with still another retrospective gesture. "At last," he writes in the final paragraph, using that formulation for the last time, "we had seen it; and we could not unsee it. We could not not have been in the Tropics" (*AL* 401).[54] The double negative in both sentences – Kingsley could "not unsee" the tropics, could "not not have been in" them – insists upon the irreversibility of tropical encounter, its permanence. At the same time, by virtue of its temporal point of vantage after the fact, that double negative cordons the tropics off from the present as surely as the fence in the frontispiece seals off the palms that stand behind it. Impossible fully to leave behind, to unsee, not not to have been in, the tropics nonetheless remain uninhabited, their potential unexhausted and therefore inexhaustible.[55] As in the passage I discuss at the beginning of this chapter, here at the end of *At Last* Kingsley confesses to a longing no consummation can put to rest: "The hunger for travel had been aroused – above all, for travel westward – and would not be satisfied" (*AL* 386).

Like Seeley's *The Expansion of England* and Kingsley's own *Westward Ho!* – along with G. A. Henty's *Under Drake's Flag: A Tale of the Spanish Main*, James Rodway's *The West Indies and the Spanish Main*, Trollope's *The West Indies and the Spanish Main*, and many similar productions – *At Last*, in varying and sometimes contradictory ways, formulates an imperial future for Britain by way of an analeptic turn to the past. In the lost world of Caribbean colonization and struggle with Spain along the coasts of Central and South America, Kingsley and other proponents of an expanded England or Greater Britain find a fitting harbinger of Britain's nineteenth-century imperial aspirations and so, at a time when very little of the Americas was possessed by the British, establish Latin America as the spectral entity at the heart of the empire. For Kingsley, such a place was not simply spectral in itself but also perhaps habitable only by specters. J. A. Froude suggests as much in his own travel narrative, *The English in the West Indies*, when he writes of Kingsley: "His memory is cherished in the island [Trinidad] as of some singular and beautiful presence which still hovers about the scenes which so delighted him in the closing evening of his own life."[56] Kingsley finds a home in Greater Britain at last, but he does so at the price of having to become a memory – yet another of those ghosts that haunt the English traveler to the islands.

4

By enlisting history and memory in the service of present expansion, Kingsley's recollected empire displaces vitality to the past. Like Darwin's and Wallace's "savages," Kingsley's precursors in the Americas mark the space and time of possibilities now lost. But the differences here outnumber the similarities. Darwin must recall savages despite an aversion to doing so, for they provide evidence of the fact of human evolution even as they stand for what is, for him, most objectionable about that fact. Wallace's memories of savages constitute a source for his critique of so-called civilization as well as a model of a more just – indeed, precisely a more human – society; for Wallace, the "savage" past provides the image of a utopian future. Kingsley, insistently looking back, rewrites the present and the future as spectral. His second-hand memories reveal the past as a locus of plenitude from which he, however, is permanently excluded. Throughout this chapter I have worked to demonstrate the implications of this particular version of savage mnemonics for empire. But I wish to conclude with a consideration of its implications for understanding Kingsley's response to another of my chief concerns in this book, and one of Kingsley's chief concerns as well: evolutionary theory.

It should be noted that Kingsley embraced that theory from the outset. On November 18, 1859, he wrote to thank Darwin for having a copy of *On the Origin of Species* (1859) sent to him; to convey his initial reaction to the book's argument, which he says he is inclined to credit; and to venture his thoughts on its implications for believers such as himself. "I have gradually learned to see," he writes,

that it is just as noble a conception of Deity, to believe that he created primal forms capable of self development into all forms needful pro tempore & pro loco, as to believe that He required a fresh act of intervention to supply the lacunas wh[ich] he himself had made. I question whether the former be not the loftier thought.[57]

Like relatively few others at the time (the *Origin* was so newly published that Darwin received his own copy only a couple of weeks before Kingsley's letter), Kingsley demonstrates an immediate willingness to accept Darwin's account of the history of life.[58] Rather than contesting that account, he sets out to incorporate it into the matrix of Christian belief and quickly arrives at the formulation that would become his standard line: that the theory of evolution by means of natural selection heightens rather than diminishes the glory of and homage due to the Christian deity.

In his children's book *The Water-Babies* (1863), an allegory of Christian redemption that borrows heavily from the picture of nature given in the *Origin*, he ventures a more elaborate marriage of his religious teachings with a natural world governed by evolutionary laws.[59] In an essay titled "The Natural Theology of the Future" (1871), he asks rhetorically: "What would the natural theologian have to say, were the first theory [the theory of evolution] true, save that God's works are even more wonderful than he always believed them to be?"[60] In *At Last* he speculates similarly: "He who makes all things make themselves may have used those very processes of variation and natural selection"; were this the case, it should result in the realization that, as he had assured Darwin, "I always knew that God was great: and I am not surprised to find Him greater than I thought Him" (*AL* 310, 246).[61] And so on: from his first reading of the *Origin* to the end of his life, Kingsley endorsed evolution.

Perhaps the boldest instance of such endorsement is one Kingsley reports to Darwin in a letter dated January 31, 1862. He writes Darwin again to inform him that he, Kingsley, had recently championed the theory of evolution by means of natural selection in the presence of the Duke of Argyll, then President of the Royal Society, and Bishop Samuel Wilberforce, the anti-evolutionist who figured famously in an 1860 Oxford debate with Thomas Henry Huxley.[62] The occasion for Kingsley's defense arose in connection with a pair of doves that had been shot, one of which seemed to present characteristics intermediate between two well-known species and so to defy identification. "My own view is," writes Kingsley, "– & I coolly stated it, fearless of consequences – that the specimen before me was only to be explained on your theory, & that Cushat, Stock doves & Blue Rock, had been once all one species."[63]

But doves, despite their potential importance to proving Darwin's theory – witness the number of pages devoted to rock doves and their descendants, domesticated pigeons, in the *Origin* – were not the only evolutionary topic on Kingsley's mind. Along with everyone else, he was deeply interested in another matter hinted at but not yet publicly addressed by Darwin: human evolution. In this letter he describes his particular worry as having to do with the difficulty posed by the "great gulf between the quadrumana [apes] & man; & the absence of any record of species intermediate between man & the ape" – that is, with what was more popularly known as the problem of the missing link. Kingsley opines, however, that this difficulty may easily be overcome, although he finds a solution neither in fuller investigation of the fossil record nor in the study of present-day "savages" mandated by the comparative method. The answer to the question of the missing link is to be arrived at, he proposes, via comparative mythology:

It has come home to me with much force, that while *we* deny the existence of any such [intermediate species between humans and other living primates], the legends of most nations are full of them. Fauns, Satyrs, Inui, Elves, Dwarfs – we call them one minute mythological personages, the next conquered inferior races – & ignore the broad fact, that they are always represented as more bestial than man … The Hounuman, monkey God of India, & his monkey armies, who take part with the Brahminae invaders, are now supposed to be a slave negro race, who joined the new Conquerors against their old masters. To me they point to some similar semi-human race. That such creatures shd. have become divine, when they became rare, & a fetish worship paid to them – as happened in *all* the cases I have mentioned, is consonant with history – & is perhaps the only explanation of fetish-worship. The fear of a terrible, brutal, & mysterious creature, still lingering in the forests. That they should have died out, by simple natural selection, before the superior white race, you & I can easily understand.[64]

As he had done in the case of the enigmatic dove, Kingsley, in a recognizably Darwinian manner, exploits the hermeneutic potential of evolutionary theory to transform an apparent problem for that theory into the basis for more evidence in its favor. In a first move, the "missing link" is identified by means of the supposition that an actual biological referent once existed for the elusive half-human creatures pervasive in myth and legend. Then, in a somewhat circular second move, not only does that supposition provide proof of the existence of such an intermediate form, it is also itself explained by Darwin's theory. One need only assume "simple natural selection," Kingsley theorizes, to account for the rarity and eventual disappearance of the supposed intermediate species, driven to extinction by the "superior white race." Kingsley does not quote Darwin but might well have: he seems to be recalling the final chapter of the *Origin*, "Recapitulation and Conclusion," which tells of "new and improved varieties" pushing out old, success breeding success.[65]

This is the letter that received the reply I discuss in chapter 1. "That is a grand & almost awful question on the genealogy of man to which you allude," Darwin responded on February 6, 1862.

It is not so awful & difficult to me, as it seems to be most, partly from familiarity and partly, I think, from having seen a good many Barbarians. I declare, the thought, when I first saw in T[ierra]. del Fuego a naked painted, shivering hideous savage, that my ancestors must have been somewhat similar beings, was at that time as revolting to me, nay more revolting than my present belief that an incomparably more remote ancestor was a hairy beast. Monkeys have downright good hearts, at least sometimes, as I could show, if I had space.[66]

Formulating the elaborate negotiation with which he would close *The Descent of Man, and Selection in Relation to Sex* (1871), Darwin assures Kingsley that to be descended from an ape is less repellant than to be descended from a savage. But the remarkable thing is how deftly he sidesteps, whether out of politeness or sheer bewilderment, Kingsley's proposition that mythology holds the key to human prehistory, his explanation of fetishism, his racist conflation of the human as such with "the superior white race" – in fact, his letter almost in its entirety.[67] Undoubtedly pleased that a well-known author and clergyman had spoken in support of his cause in the presence of men with such institutional authority as the Duke of Argyll and Bishop Wilberforce, Darwin nonetheless seems to have found Kingsley's attempt at removing an impediment to the acceptance of human evolution untenable or uninteresting – despite the fact that Huxley, only a year later, would open *Evidence as to Man's Place in Nature* (1863) with a remarkably similar gambit. "Ancient traditions," he writes, "when tested by the severe processes of modern investigation, commonly enough fade away into mere dreams: but it is singular how often the dream turns out to have been a half-waking one, presaging a reality." He goes on to echo Kingsley still more closely, noting that although "Centaurs and Satyrs" exist "only in the realms of art, creatures approaching man more nearly than they in essential structure, and yet as thoroughly brutal as the goat's or horse's half of the mythical compound, are now not only known, but notorious."[68]

Darwin's lack of interest in Kingsley's letter may derive from the approach to the past it encodes, an approach characteristically different from Darwin's. To begin with, and here like Darwin, Kingsley imagines the past to be the source of answers to present-day questions. But the past to which he turns is not the tangible one of the fossil record or archeological remains but the intangible one of human systems of belief, which, in the case of the "missing link," he interprets as collective memories of once-extant creatures, memories sedimented and passed down in the form of myth. Presented as evidence of the existence of a form intermediate between modern humans and proto-hominids, that myth also bears witness to a lack in the present. The "terrible, brutal, & mysterious creature" to which it attests is in Kingsley's rendering an ill-adapted or inferior version of the human. But it also stands as the lingering vestige of lost powers. Whence its god-like status in the eyes of those who have supplanted it: "That such creatures shd. have become divine, when they became rare, & a fetish worship paid to them … is consonant with history – & is perhaps the only explanation of fetish-worship." The extinction of this "semi-human race" marks at once the advent of the human as such and the disappearance of a "bestial" power.

Like the species past about which he writes to Darwin, the national-imperial past that haunts Kingsley in the pages of *At Last* survives to attest to how the modern came into being as well as to what was lost in the process. The English dead, the tragedies and romances of the Americas, explorers and buccaneers: they remain in memory to remind Kingsley and his readers of what once was. Unlike elves, dwarves, and satyrs of legend, however, these predecessors bear witness to a time and place Kingsley wishes to make live again. But the land of promise is as extinct as the missing link. Thus it is that, in his own terms, Kingsley may be said to fetishize the past, worshiping it precisely because it cannot be brought back.

Darwin, who looks to still-extant indigenous peoples as evidence of the missing link, nonetheless finds the notion of his kinship with them, as he writes Kingsley, "more revolting than [his] present belief that an incomparably more remote ancestor was a hairy beast." For him, the savage past is so proximate, so similar to the "civilized" present that it constitutes a threat here coded as visceral revulsion. Kingsley, by contrast, depicts savagery as utterly distinct from himself. When arguing in *At Last* that British rule has been a boon to Trinidad, for instance, he credits it with saving the island from having to endure the "massacre of the respectable folk, the expulsion of capital, and the establishment (with a pronuncia-mento and a revolution every few years) of a republic such as those of Spanish America, combining every vice of civilization with every vice of savagery" (*AL* 107).[69] In the book's final chapter, the same savagery stands in the way of visiting South America. When the material represented by ellipses in my initial quotation of it has been replaced, the passage recounting the westward fever that overtook Kingsley as he turned away from the Americas and back toward Britain reads as follows:

The hunger for travel had been aroused – above all for travel westward – and would not be satisfied. Up the Oroonoco we longed to go: but could not. To La Guayra and Caraccas we longed to go: but dared not. Thanks to Spanish Republican barbarism, the only regular communication with that once magnifi-cent capital of Northern Venezuela was by a filthy steamer, the Regos Ferreos, which had become, from her very looks, a byword in the port. On board of her some friends of ours had lately been glad to sleep in a dog-hutch on deck, to escape the filth and vermin of the berths; and went hungry for want of decent food. Caraccas itself was going through one of its periodic revolutions – it has not got through the fever fit yet – and neither life nor property were safe.

But the longing to go westward was on us nevertheless. (*AL* 387–88)

Echoing arguments for empire promulgated in Seeley's *The Expansion of England* and elsewhere, Kingsley posits imperial administration in the Americas as the antidote to the chaos and violence of democratic self-rule. But if what he describes as the destructive savagery of the South American republics is safely distant, so, too, are the glorious deeds of the founders of the British Empire. Inhabiting a kind of perpetual afterlife, Kingsley is as little threatened by the notion of his kinship with the savage as he is given hope by the possibility of resettling the Americas.

Whence the ugly conclusion to Kingsley's letter of January 31, 1862:

I hope that you will not think me dreaming – To me, it seems strange that we are to deny that any Creatures intermediate between man & the ape ever existed, while our forefathers of every race, assure us that they did – As for having no historic evidence of them – How can you have historic evidence in pre-historic times? Our race was strong enough to kill them out while it was yet savage – We are not niggers, who can exist till the 19th century with gorillas a few miles off.[70]

On one hand, Kingsley denigrates Africans for apparently being so unevolved as not to have exterminated "any Creatures intermediate between man & the ape." On the other, the presumed superiority of his and Darwin's "race" inheres in its own primitive vitality: it "was strong enough to kill them out while it was yet savage." Present-day Britons are defined as lacking savagery, but that lack is not only to be celebrated. To recall his predecessors in the Americas is for Kingsley to be reminded of what was once possible as well as of what is possible no more. Exuberant and powerful, those predecessors and the imperial aspirations for which they stand are also – like satyrs and dwarves, like palms in botanical gardens – walled off from the present, to be worshiped because, although they can be remembered, they cannot be reached.

And so Kingsley, just as he assimilates evolutionary theory into his Christian belief, assimilates it into his formulation of relations among past, present, and future. It does not challenge that formulation so much as provide yet another way to confirm it. Consider, as a final instance, a passage from his pre-Darwinian work of popular natural history, *Glaucus; or, The Wonders of the Shore* (1855). Taking his reader on an imagined geologizing expedition, Kingsley writes of a smooth area of rock:

> It was the crawling of a glacier which polished that rock-face … Aeons and aeons ago … those marks were there; the records of the "Age of ice"; slight truly; to be effaced by the next farmer who needs to build a wall; but unmistakable, boundless in significance, like Crusoe's one savage footprint on the sea-shore: and the naturalist acknowledges the finger-mark of God, and wonders, and worships.[71]

The present bears evidence of the past. Although tenuous and fragile, destined to disappear, that evidence testifies to a time when "the finger-mark of God" was placed on the earth, a time that therefore compels both wonder and worship. Already at one remove from the present, such a time is distanced yet again by way of simile. Likening remains of the glacial past to "Crusoe's one savage footprint," Kingsley renders the past textual even as he ties its boundless significance to savagery and, not incidentally, to the Americas. No westward journey, however protracted, could arrive at this destination. The longing for such a journey carries Kingsley toward distant shores, but they are only – they are always – the shores of memory.

CHAPTER 4

W. H. Hudson's memory of loss

Edward Garnett opens an obituary tribute to W. H. Hudson with a description of his friend as he appeared not long after he had died:

When after Hudson's death I looked down upon his face as he lay on his bed in the shadowed room, I saw before me the calm death mask of a strong chieftain. All the chiselled, wavy lines of his wide brow, the brooding mournfulness and glowing fire of his face had been smoothed out. He was lying like some old chief of the Bronze Age, who, through long years of good and ill, had led his tribe. And now for him only remained the ancient rites, the purging fire, the cairn on the hillside, and the eternity of the stars, the wind, the sun.[1]

The place is London; the time, August 1922. But the picture Garnett paints looks back to prehistory, to an archaic world of "tribe," "ancient rites," "purging fire," and a "cairn on the hillside." Conjuring up what he takes to be the definitive elements of that world, Garnett renders his friend, in death as well as in life, a human archaism. Hudson himself could not have crafted a more characteristic portrayal. In fact, despite having been written posthumously and appearing over another man's name, this rendition of Hudson as a Bronze Age chieftain is in some sense a self-portrait. It distills into a single image many similar representations he made in conversation, in letters to friends and admirers, and in his books. He was a "red man," he declared, a "wild man of the woods," an ancient survival fated to endure a life of temporal exile in the modern age.[2] He was, in short, a savage.

Hudson's deep and enduring identification with savagery derived from the events of his own life and those of world history as he believed it to be unfolding. Born on an estancia near Buenos Aires in 1841, the son of émigrés from the United States, he was a passionate observer of South American flora and fauna throughout his childhood and youth. I write "observer," but the connotation of detachment the word carries almost renders it false when applied to Hudson, for whom the natural world was not a separate or separable entity but an extension of himself, animated

121

with a significance "so powerful," he reports in his autobiography, "that I am almost afraid to say how deeply I was moved by it."[3] In 1874 he emigrated to England, where he remained until the death about which Garnett writes so evocatively.

As emplotted by most of his biographers, the nearly half century Hudson lived away from the land of his birth was occupied by a long struggle against poverty and obscurity crowned with triumphant success. The estimate of his contemporaries well indicates the extent of the success. Garnett, at a later moment in his tribute, calls Hudson "our greatest English naturalist," crediting him with "enter[ing] more intimately and steep[ing] himself more intensely in the protean spectacle of Nature's life than have any of his rivals, Bates or Wallace or Belt or Jefferies or Burroughs."[4] Alfred Russel Wallace himself was of the same opinion. In an interview late in life he expresses the judgement that Hudson's *The Naturalist in La Plata* (1892) "was a finer work than Darwin's 'Voyage of the Beagle' or my own 'Malay Archipelago.'"[5] Ford Madox Ford goes further, opining that at the beginning of the twentieth century "there was no one – no writer – who did not acknowledge without question that this composed giant was the greatest living writer of English."[6] Despite such acclaim and the relative material prosperity that came with it, Hudson himself often spoke of the years he spent in England as a posthumous existence.[7] To be cut off from South America, the originary site of his fierce love for the natural world, was to him a loss so consequential as to constitute a kind of death.[8]

That loss, bitter enough in itself, was made more so by the course of events in the land he had left behind. Half a decade after his departure from Argentina, a brutal war of extermination waged by the Argentine army against so-called "Indians" opened the pampas to widespread European and creole settlement and cultivation.[9] Following hard on the heels of this decimation of the indigenous population, the resulting destruction of the grasslands they and Hudson had called home transformed what had been merely personal displacement into a state at once ubiquitous and permanent. As Hudson tells it, neither he nor anyone else could return to the world of his youth – not, to cite the usual truisms, because youthful enthusiasm fades or maturity puts an end to childhood innocence but for the starkly literal reason that there was nothing any longer to return to. The vast expanse of open land known in Spanish as *la pampa*, along with its characteristic animals, and especially the native avian and mammalian life he loved, had disappeared or were fast disappearing. Hudson knew, moreover, that a similar sequence of events was unfolding across the globe.

So it happens that, as Ian Duncan writes, "Hudson's sense of loss ... was mythic, for he could understand his life as having traced a universal disaster – the fall of nature under the onslaught of civilization."[10] He sought to stave off what additional destruction he could by engaging in political activism, including a letter-writing campaign against the use of feathers in hats and clothing and, most notably, a key role in the founding of what was to become the Royal Society for the Protection of Birds, an early, important, and still-extant wildlife conservation organization.[11] Whatever success such efforts might achieve, however, was powerless to reverse the catastrophe that had already occurred. What had been lost, he insisted, was lost irretrievably. And so, with restoration impossible, Hudson, as a man who had seen Paradise before the fall, felt his duty to be that of a memorian.

I borrow this term from Andreas Huyssen, who, in *Present Pasts: Urban Palimpsests and the Politics of Memory* (2003), invokes it to name an investment in the past as memory rather than history. Whereas historians, in Huyssen's account, seek out that version of the past the accuracy of which may be verified by appeal to documents, memorians draw on recollection to piece together the varied and particular memory-scapes that documents fail to register or preserve.[12] To call Hudson a memorian is thus to distinguish him from a historian but also, in the context of the other figures I treat in this book, from a natural historian, an evolutionary theorist, or a writer about what were, for most Britons, exotic lands. He was all of these at once, and more besides. But history, the natural world, evolutionary biology, South America: all, for him, found their meaning in memory.

In what follows, I develop the implications of that claim by way of a reading of three of Hudson's major works of natural history: *The Naturalist in La Plata*, *Idle Days in Patagonia* (1893), and *A Hind in Richmond Park* (1922). These books are among those categorized in the front matter of the first American edition of *A Hind in Richmond Park* as "Reminiscences of a Naturalist."[13] Accordingly, the stakes involved in my attention to them may be indicated by invoking another aspect of Huyssen's argument in *Present Pasts*: his discussion of what he calls the "memory fever" that has infected both the developed and the developing worlds during the past few decades, a compulsion to plant memory gardens, erect memorials, transcribe oral histories, and otherwise bear witness to a past thought to be quickly receding from view. For Huyssen, this recent and widespread turn to memory results directly from the impact of the new media, which, because of their ability to capture and permanently archive vast amounts of information, paradoxically threaten an overwhelming amnesia. "[T]he relationship between memory and forgetting," he writes, is "actually being transformed under cultural pressures in which new information technologies, media politics, and fast-paced consumption are beginning to take their toll."[14] In response, he claims, "we are trying to counteract [the] fear and danger of forgetting with survival strategies of public and private memorialization." Explicitly contrasting the urgency of this present-day sense of the historical and political consequentiality of memory with what he sees as the narrower conception of its import during the nineteenth and early twentieth centuries, he asserts that "neither Wordsworth nor Proust was compelled to think about memory and forgetting as social and political issues of global proportions, as we are today."[15]

Whatever one thinks of the soundness of that observation as it relates to the great nineteenth-century poet and great twentieth-century novelist of memory, it seems indisputable that Hudson's memorializing imperatives evidence a role for memory just as political in intent and global in implication as, to take one of Huyssen's chief examples, Christo's 1995 wrapping of the Reichstag. Further, if Huyssen's chronology admits of modification, so, too, does his causality. For, somewhat startlingly, Hudson reveals European colonialism to be a precursor of and analogue for the new media insofar as colonialism's musealization of nature, and of indigenous peoples and cultures, went hand in hand with its extension of the reach of ecological and human disaster. Hudson's own febrile reminiscences come into being as a response to that disaster, so that his vocation as a memorian may be more fully characterized by appeal to Cathy Caruth's explication of "trauma memory." "To be traumatized," she avers, "is precisely to be possessed by an image or event."[16] In styling himself a savage, Hudson meant specifically that he was a child of the wilderness who had outlived it – and whose recollections contained its only remaining traces. An Ancient Mariner who had roamed an inland sea that no longer existed, the wide-open pampas, he was compelled again and again to draw on memory in the service of relating how things used to be.

This is the keynote of Hudson's work: the solemn invocation of a lost world as a form of testimony, protest, and lament. His memories of that world often provide glimpses of plenitude, realizing in imagination and on the page images of wild beauty. But they cannot do so without simultaneously insisting on past, present, and future catastrophe.[17] This among other things distinguishes the structure of feeling in Hudson's corpus from what might appear to be unmitigated or self-indulgent nostalgia. Even his titles say as much. He called his first novel, a collection of picaresque adventures and gaucho tales set in Uruguay, *The Purple Land that England Lost* (1885); a children's book he wrote, *A Little Boy Lost* (1905); his autobiography, *Far Away and Long Ago* (1918). That autobiography's most telling line of chapter description reads simply "My boyhood ends in disaster" (*FA* xi). And a sentence in a description of one of the chapters in the last book he wrote, *A Hind in Richmond Park*, declares: "The beautiful has vanished and returns not."[18] The specific reference is to the many species of birds driven to extinction during Hudson's lifetime, but the epigrammatic formulation seeks to name a more general condition: the worldwide destruction of the beauty and complexity in and of nature, which, if it survives at all in its intact form, does so only in memory. To tell of that memory is, therefore, always also to tell of irreparable damage. Neither for Hudson nor for anyone else is there any genuine or full going back, not even in recollection. "The beautiful has vanished and returns not" – nor can we ever return to it.

Hudson thus realizes and plays out the consequences of an unforeseen implication of evolutionary theory. Many discussions of the cultural resonance of evolution, both during the Victorian era and at present, stress the threat it poses insofar as it connects humans intimately to what must now be called other animals, and hence to the pre- or inhuman. Charles Darwin's rhetorical efforts at the end of the *Descent of Man, and Selection in Relation to Sex* (1871) directly invoke and attempt to allay that threat: "He who has seen a savage in his native land will not feel much shame, if forced to acknowledge that the blood of some more humble creature flows in his veins."[19] Hudson himself adduces such shame as one of the reasons for resistance to Darwinism. "But we know that the fact of evolution in the organic world was repellent to us," he writes in *A Hind in Richmond Park*, "because we did not like to believe that we had been fashioned, mentally and physically, out of the same clay as the lower animals" (*HRP* 276). But he also perceives that evolution, even as it forges a link with the pre-human and thus with the animal and the archaic, also posits a break from the past, a break absolute and therefore, to him, terrible.

In this respect evolutionary theory as developed by Darwin and Wallace differs markedly from previous explanations of the origins and workings of the biological world. The doctrine of special creation views that world as subject to ongoing renewal, and none of its corollaries definitively excludes the possible re-creation of a formerly numerous but now extinct species. Pre-Darwinian versions of transmutation theory such as Lamarck's, despite its suggestion of an inherent organismal tendency to progress and thus to change over time, nonetheless imagines past, present, and future as a seamless continuity. "This reading of the past," Gillian Beer observes, "has no need of the concept of usurpation."[20] But the theory of natural selection insists on nearly infinite complexity and thus on unrepeatability: each species, the product of innumerable tiny, random variations accumulated over the course of millions of years, is unique and cannot appear twice.[21] One of the hardest lessons of evolution, this insistence informs Hudson's wonder at the beauty of various organisms and the deep despair of his lament for their loss.[22] Because many of those organisms were bound up so intimately with his own past, it also informs his sense of being an exile – on view in his autobiography and his natural history writings as well as in his novels, whose protagonists, from Richard Lamb of *The Purple Land that England Lost* to Mr. Abel of *Green Mansions* (1904), are themselves exiles from their native lands. It is as if Hudson were the first clearly to perceive that the doctrine of evolution mandates an ineluctable version of leave-taking without return, a complete and final disappearance.

But Hudson also saw that evolutionary theory, having set the problem by making thinkable destruction without possibility of restoration, provided a solution as well. Marjorie Grene and David Depew note in *The Philosophy of Biology*: "After Darwin, the whole multitudinous and variegated biota on and over and below this earth forms a family, not nearer to or farther from some social ideal type, but related to one another through generation, just as our families are – though ... in a much more complex pattern, of which we perceive only scattered bits."[23] By asserting as its chief insight the fact of the relatedness, by descent, of all life on the planet, evolution promises to make the otherwise inaccessible past available by locating it within each living organism – and so within Hudson as well as every other human. Whence the central role in Hudson's thought played by his insistence that he was a savage. The hosts of species replaced by others in the course of evolutionary change or driven to extinction by human activity leave behind, along with their bones, memories that may be summoned at need but that remain memories only – recollections of a vanished past rather than that past itself. But savages, if in actuality dying out or in the process of being massacred or at the least driven further and further from the haunts of the civilized, escape extinction for Hudson by residing, in essence, inside the "modern" human. Such a notion is transparently self-serving, rendering the extermination of indigenous peoples bearable by suggesting their disappearance will not impoverish the world as long as those who remain may "go native" by going within.[24] But this ideologically risky identification with savagery also tells us something important about loss, the memory of loss, and memory as loss. For if in one sense turning within and to the past in the face of present-day loss constitutes an admission of defeat, in another, via a surprising reversal, the savage past available to recollection may be enlisted in the service of the restoration of a public, collective memory of wildness and savagery harnessed to political ends.

I

Hudson begins *The Naturalist in La Plata*, the book that made him famous and secured his career as a writer, by decrying the widespread ecological damage inflicted by European colonization. "We hear most frequently of North America, New Zealand, and Australia in this connection," he writes in the book's first chapter, "The Desert Pampas," "but nowhere on the globe has civilization 'written strange defeatures' more markedly than on that great area of level country called by English writers *the pampas*."[25] The sentence conveys Hudson's sense of the worldwide scale of environmental disaster as well as the special status of the pampas, in his view the region that has suffered most. He frames the book to follow as a description, "from the field naturalist's point of view," of the pampas of his childhood and youth, a description that will take on the contours of a memorial to the thing described: "In view of this wave of change now rapidly sweeping away the old order, with whatever beauty and grace it possessed, it might not seem inopportune at the present moment to give a rapid sketch … of the great plain, as it existed before the agencies introduced by European colonists had done their work, and as it still exists in its remoter parts" (*NLP* 4). In the space of a few pages, Hudson establishes *The Naturalist in La Plata* as a distinctive sort of natural history: one drawn from memory rather than transcribed during direct observation; occasioned by the loss of the things remembered; and, therefore, inflected throughout with a tone of lament.

The claim that "nowhere on the globe has civilization 'written strange defeatures' more markedly" than on the pampas singles out the land of his birth as exceptional but also makes it representative. In a characteristic gesture, Hudson adumbrates a class of losses by focusing on the one he perceives to be the most egregious. The brief, unattributed quotation is from Shakespeare's *Comedy of Errors*, and it resounds in several ways at once, providing insight into Hudson's relation not only to the past, and to South America and Britain, but also to the natural world as such. Most evidently, it constitutes one instance among many in Hudson's work of an anxiety not so much of influence as of affiliation and expertise. In *At Last: A Christmas in the West Indies* (1871), as I demonstrate in the previous chapter, Charles Kingsley valorizes the past and embraces his own status as a latecomer, compulsively citing those writers about and travelers to the Americas who preceded him. Hudson repeats while at the same time reversing Kingsley's geography of belatedness: an Argentine by birth but a Briton by choice and ancestry, his apparently offhanded allusion to Shakespeare, arguably preeminent among literary icons of Englishness, serves to demonstrate thorough knowledge of and belonging to his adopted home. Elsewhere Hudson positions himself as an amateur naturalist in a world of professional biologists, confirming his right to belong by displaying exhaustive familiarity with the latest scientific findings as well as indulging in a frequently ill-advised willingness to challenge those with more institutional authority than he himself possessed. Among his earliest publications, for instance, is an attack on Darwin's description of a South American woodpecker living in the treeless pampas, given in *On the Origin of Species* (1859) as an example of a fact impossible to explain on the theory of special creation.[26] Here, with the quotation from Shakespeare, at the moment Hudson establishes his right to speak for the vanishing South American wilderness, he establishes, too, his right to speak as a European and specifically a Briton.

But the phrase from *The Comedy of Errors* also and perhaps more importantly functions to metaphorize European spoliation of the natural world as the effects of sorrow on the human countenance. In the play, Antipholus of Ephesus, on seeing his father, Aegeon, after a seven-year separation, fails to recognize him – a failure that Aegeon attributes to his own altered appearance: "O, grief hath chang'd me since you saw me last, / And careful hours, with Time's deformed hand, / Have written strange defeatures in my face."[27] "[G]rief," "careful hours," and the "deformed hand" of time become, in Hudson's rendering, simply "civilization"; the "face" that has been scarred or defeatured by civilization's writing, nature. For Darwin in the *Origin*, the "face of nature bright with gladness" presents a deceptive appearance that the deliberate recollection of predation and competition must work to overcome if evolution by natural selection is to be understood.[28] For Hudson, the face of nature only too obviously bears marks of destruction. Those marks are not, however, signs of internecine struggle but "strange defeatures" inscribed by human, and specifically civilized, hands. Recollection figures, accordingly, not as a reminder of unacknowledged or forgotten struggle but as a means to restore in imagination an aboriginal world:

the Argentine pampas "before the agencies introduced by European colonists had done their work" (*NLP* 4).

Much of the remainder of the first chapter of *The Naturalist in La Plata* lovingly memorializes that prelapsarian pampas – its endemic grasses and flowers, rodents, cats, dogs, marsupials, and, above all, birds. Calling the pampas "my 'parish of Selborne,'" and so, in another moment of carefully articulated relation to predecessors, identifying himself with the proto-typical English field naturalist and natural history writer, Gilbert White, Hudson adopts a celebratory tone as he moves from creature to creature as if he were recounting a litany of miracles (*NLP* 5).[29] But celebration comes to an abrupt end when, after an admiring description of the rhea, the "grand archaic ostrich of America," he introduces a catalogue of those species in the process of being extirpated (*NLP* 26). The rhea, once common, is disappearing.

And with the rhea go the flamingo, antique and splendid; and the swans in their bridal plumage; and the rufous tinamou – sweet and mournful melodist of the eventide; and the noble crested screamer, that clarion-voiced watch-bird of the night in the wilderness. These, and the other large avians, together with the finest of the mammalians, will shortly be lost to the pampas utterly as the great bustard is to England, and as the wild turkey and bison and many other species will be lost to North America. What a wail there would be in the world if a sudden destruction were to fall on the accumulated art-treasures of the National Gallery, and the marbles in the British Museum, and the contents of the King's Library – the old prints and mediaeval illuminations! And these are only the work of human hands and brains … – and man has the long day of life before him in which to do again things like these, and better than these, if there is any truth in evolution. But the forms of life in the two higher vertebrate classes are Nature's most perfect work; and the life of even a single species is of incalculably greater value to mankind, for what it teaches and would continue to teach, than all the chiseled marbles and painted canvases the world contains … Like immortal flowers they have drifted down to us on the ocean of time, and their strangeness and beauty bring to our imaginations a dream and a picture of that unknown world, immeasurably far removed, where man was not: and when they perish, something of the gladness goes out from nature, and the sunshine loses something of its brightness. (*NLP* 28–30)

These lines function at once as an incantatory tribute to beloved animals and a prediction of overwhelming catastrophe. Inverting the trajectory of the opening paragraphs of "The Desert Pampas," which begin with global disaster but quickly focus in on the pampas, this passage expands its purview from South American to British and North American extinction – as well as from a single species, the rhea, to all "the forms of life in the two higher vertebrate classes." The globe in its entirety is re-established as the appropriate scale on which to measure what is in the process of disappearing. Finally, and in its way most tellingly, an additional measure of loss is introduced with the assertion that any one among the many species of birds and mammals being destroyed is more precious than the entirety of "the accumulated art-treasures of the National Gallery, and the marbles in the British Museum, and the contents of the King's Library." Having demonstrated, by way of references to Shakespeare and Gilbert White, his intimate knowledge of Britain's heritage, Hudson, as a means to cast into relief the enormity of ecological disaster, asks readers to contemplate the obliteration of the contents of three of its most revered cultural institutions. The outcry at such an event would doubtless be staggering; how much louder and more anguished then, he insists, should the response to mass extinction be; how much worse it is to lose species than books and artworks.

Evolution, although named in the passage only once and in an aside, plays a defining role. The plangent note of irredeemable loss suffusing the whole is underwritten by Hudson's reminder of the vast "ocean of time" required for the process that results in the production of each species, as well as of the unrepeatability of that process and hence the absolute finality of extinction. Evolution points the contrast between animals on one hand and cultural artifacts on the other insofar as the latter may be created quickly and, in the event that they are destroyed, recreated. Humans in this sense stand apart from the natural world because their creations are comparatively easy of accomplishment. Further, because the things humans make may be made again, they are in a sense indestructible. Read in this light, Hudson's deployment of evolution works to separate humans from the rest of nature. But considered in another light the role of evolution places humans firmly back in the natural world. The effect of the imagined destruction of inherited literary and artistic treasures would be mitigated, Hudson suggests, because future productions will surely be "better" than those destroyed. The qualification to this claim, "if there is any truth in evolution," assumes humans to be as susceptible to evolutionary pressures as other animals – and assumes, too, that their cultural products, like any other sort of specialized capacity, will be rendered better, more "fit," because of it.

Human separation returns at the end of the passage, however, to the extent that the human future remains open, subject to the same progressive improvement that has produced its past and present, while human activity itself curtails the future of birds and nonhuman mammals. That distinction is driven home on the chapter's final page, where Hudson takes the measure of human progress by envisioning the attitude a future civilization will hold toward his own day. What had been retrospection becomes prophecy as he looks far forward in time rather than back to the recent past. But such is the pervasive effect of memory in his thought that even prolepsis must be expressed by way of the language of recollection:

It is hardly to be supposed or hoped that posterity will be satisfied with our monographs of extinct species, and the few crumbling bones and faded feathers, which may possibly survive half a dozen centuries in some happily-placed museum. On the contrary, such dreary mementoes will only serve to remind them of their loss; and if they remember us at all, it will only be to hate our memory, and our age – this enlightened, scientific, humanitarian age, which should have for a motto "Let us slay all noble and beautiful things, for tomorrow we die." (*NLP* 31)

Foreseeing a time when wild animals exist only in the form of mementoes that serve as reminders of loss, and when, as a result, the human past exists only as a memory that feeds hatred of that which is remembered, Hudson folds past, present, and future into a seamless and apparently inescapable unnatural history of destruction.[30] Despite its reference to the passage of "half a dozen centuries," however, this description suggests that, for Hudson, the future is now. For he writes in and of a time when "monographs of extinct species" and "crumbling bones and faded feathers" already attest to a world that no longer exists, when many species continue to live only in the form of what is remembered about them – the subject, after all, of *The Natural History of La Plata* in its entirety. In its first chapter, from the opening invocation of "strange defeatures" to the final evocation of the hateful memories of the present likely to be held by humans living ages hence, Hudson casts the book as a memorial to a nature no longer extant: recollected rather than collected, remembered rather than seen.

But even as "The Desert Pampas" establishes the content of the book to follow in its entirety as drawn from recollection, it also centers on a specific memory that will serve, Hudson announces, as representative of all the others: that of "the last occasion on which [he] saw the pampas grass in its full beauty" (*NLP* 7). Despite this specification, what Hudson presents bears only tangentially on vegetation or landscape. Nor does a beautiful bird or nonhuman mammal occupy the center of attention. Readers are given an account of neither "The Dying Huanaco," say, nor "The Puma, or Lion of America," to cite the titles of two subsequent chapters. Rather, although introduced as having to do with grass in flower, the memory in question turns out to be a memory of "savages."

Hudson and a companion, having ridden all day through a seemingly uninhabited wilderness, encounter "five mounted Indians" near sunset. Coming to a sudden stop, the five riders stand atop their horses' backs and begin to scan the horizon. "Satisfied that they had no intention of attacking us," Hudson writes,

> and were only looking out for strayed horses, we continued watching them for some time, as they stood gazing away over the plain in different directions, motionless and silent, like bronze men on strange horse-shaped pedestals of dark stone; so dark in their copper skins and long black hair, against the far-off ethereal sky, flushed with amber light; and at their feet and all around, the cloud of white and faintly-blushing plumes. That farewell scene was printed very vividly on my memory, but cannot be shown to another, nor could it be even if a Ruskin's pen or a Turner's pencil were mine; for the flight of the sea-mew is not more impossible to us than the power to picture forth the image of Nature in our souls, when she reveals herself in one of those "special moments" which have "special grace" in situations where her wild beauty has never been spoiled by man. (*NLP* 8)

In this, their first appearance in *The Naturalist in La Plata*, "savages" signify in ways that presage their role in the rest of the book as well as in Hudson's corpus as a whole. Apparently in keeping with their representative function, these riders are rendered as works of art. As Ezra Pound has it in his laudatory review of Hudson's works, they are "living effigies in bronze rising out of the white sea of the pampas."[31] If this sort of portrayal appears to elevate indigenous people, promoting them into an incarnation of nobility and grace, it does so by converting them entirely into an object of the gaze (even if they are themselves also gazing), stripping them of agency by casting them in metal. The dynamic is a familiar one; like the "art-treasures of the National Gallery, and the marbles in the British Museum," these "bronze men" in their "copper skins" on "pedestals of dark stone" are at once more and less than human and more and less than nature. Paradoxically, however, they also exemplify the natural world, or at least that part of it corresponding to Hudson's memory of the last time he saw the pampas "in its full beauty." Savages, by way of a detour through the artifactual, come to represent part of that natural paradise to which Hudson felt so close – and not simply any part of it but its most representative part, a synecdoche for the rest.[32] Or, to be more precise, they are a metonymy of a synecdoche: human artifacts standing in for a memory of pampas grass, which itself stands in for the lost pampas as such.

Figure 4 "Pampas Grass: Indians on the Look-out for Strayed Horses," frontispiece,
W. H. Hudson, *The Naturalist in La Plata* (London: Chapman and Hall, 1892).
Courtesy the Thomas Fisher Rare Book Library, University of Toronto.

Such complexity may account, in part, for the attitude Hudson takes
toward the possibility of representing these figures. On one hand, they
must be depicted, standing in as they do for the pampas of memory. On
the other, Hudson insists they cannot be depicted with accuracy. Neither
Hudson's Indians, serving as place-holders for a moment of heightened
closeness to and delight in the natural world, nor that moment itself can
be represented or transmitted from one person to another, despite having
been "printed very vividly on [Hudson's] memory." (The implied metaphor of photography for the workings of the memory is something to
which I will return below.) Moreover, this limit to humans' powers of
representation marks a distinction between them and the rest of the natural world: "the flight of the sea-mew is not more impossible to us than
the power to picture forth the image of Nature in our souls." But then, of
course, Hudson does "picture forth" that image, and not once but twice:
verbally in the description itself, visually in a plate that stands as frontispiece to *The Naturalist in La Plata* as a whole, over the caption "Pampas
Grass: Indians on the Look-out for Strayed Horses" (see Figure 4).[33]
Sharply "printed" on the memory, incapable of being represented but in
fact represented twice, these "Indians on the Lookout for Strayed Horses"
in both instances represent lost nature: Hudson's memory of the last time
he saw pampas grass in its full glory and, more strikingly, the whole of

the vanishing natural world that it is his vocation to recall and to mourn. Here, in the first chapter of Hudson's first book of natural history, "savages" assume the vexed contours that they will retain in all of his writings. They are not identical to Hudson but he identifies intensely with them; they are not nature but nonetheless stand in for it; they exist in a double temporal register, vibrantly alive in the moment of observation but persisting into the present only in and as memory.

Despite their centrality to "The Desert Pampas," savages play only a muted role in the rest of *The Naturalist in La Plata*, which is largely taken up with a series of chapter-long reflections on the region's distinctive animals, from "The Mephitic Skunk" and "Humming-birds" to "The Woodhewer Family." Each such chapter forms an essayistic set-piece on the animal or animals in question, featuring Hudson's personal observations and speculations as well as reference to a wide array of natural-historical authorities. Other chapters range more widely, treating phenomena such as instinct, parasites, and "Music and Dancing in Nature" (the title of chapter 19) as problems in biological theory. Although never the chief object of attention, savages put in occasional appearances to illustrate particular points. In chapter 17, for instance, "The Crested Screamer," Hudson writes of live specimens of these birds kept in the London Zoo and the melancholy effect their cries have on him as he crosses through Regent's Park. At the end of the chapter, which gives as the cause of his melancholy the conviction that the crested screamer will inevitably become extinct, Hudson notes:

It is sad to reflect that all our domestic animals have descended to us from those ancient times which we are accustomed to regard as dark or barbarous, while the effect of our modern so-called humane civilization has been purely destructive to animal life. Not one type do we rescue from the carnage going on at an ever-increasing rate over all the globe. (*NLP* 230)

Although considered to be "dark or barbarous," Hudson avers, human prehistory is in one sense less so than the era of modernity insofar as the latter is "purely destructive to animal life." Then, via an unmarked slippage, Hudson moves from ancient times to the present: "With grief and shame, even with dismay, we call to mind that our country is now a stupendous manufactory of destructive engines, which we are rapidly placing in the hands of all the savage and semi-savage peoples of the earth, thus ensuring the speedy destruction of all the finest types in the animal kingdom" (*NLP* 230).[34] Not only do moderns fail where barbarians succeeded, that is, but they provide present-day barbarians with the means to destroy.

To this point in the book, savages are generally allied with the natural world and opposed to a humanity understood as "modern" or "civilized." That alliance is occasionally qualified, as in "The Crested Screamer," where contemporary savages, armed by the civilized, participate in their wanton destruction of animal life. And the specifics of that opposition are notable to the degree that they anticipate a certain primitivism and thus counter widespread Victorian attitudes by reversing the polarities of value between savage and civilized, casting prehistoric savages in the role of creators and preservers and featuring Hudson's European contemporaries as destroyers. Like Wallace, Hudson inverts the common account of relations between savage and civilized in the service of a critique of nineteenth-century Europe, and specifically of European colonialism.

So goes *The Naturalist in La Plata* until the final chapter, "Seen and Lost," which at first seems to have nothing to say about savages but ends up with what Hudson regards as an encounter with savagery that sheds retrospective light on all the parts of the book that have come before as well as prospective light on the future shape of Hudson's savage mnemonics. For it is in this final chapter of *The Naturalist* that memory begins to offer what is to become for Hudson its salvational potential: the means to bridge the gap between "savage" and "civilized" by returning moderns to a savage state, and thus the possibility of resurrecting something of that past whose loss so preoccupied him.

The chapter opens with a parable. A lapidary is shown a beautiful and heretofore unknown kind of gemstone by someone who then immediately disappears with it. The imagined reaction of the lapidary conveys, Hudson insists, what he himself has often felt: "A feeling such as that would be is not unfrequently experienced by the field naturalist whose favoured lot is to live in a country not yet 'thoroughly worked out,' with its every wild inhabitant scientifically named, accurately described, and skillfully figured in some colossal monograph" (*NLP* 360). Having briefly seen but been unable to capture an individual of a species new to science, the naturalist registers at once the excitement of discovery and the sorrow of loss. The simile likens individuals of animal species to gems but is imperfect, Hudson points out, insofar as nature is more prolific in varieties of life than in varieties of gemstone – so that the sense of disappointment in the case of the species glimpsed but not captured is recompensed with the expectation of future sights of other novel creatures.

Despite this implication of the inexhaustibility of the natural world, the chapter's title signals that what is to follow will be a continuation of the rest of the book's lament, and it does so specifically by yoking the visual register to disappearance: "seen and lost." A subsequent passage that employs photography as a metaphor for memory solidifies the connection. If the lapidary doubts whether what he has been shown actually exists, Hudson contends, the naturalist is certain: "for there it is, the new strange form, photographed by instantaneous process on his mind, and there it will remain, a tantalizing image, its sharp lines and fresh colouring unblurred by time" (*NLP* 361). The very precision of the image, however, preserved as if by photographic capture, only serves to heighten the naturalist's sense of missed opportunity. Recall that in the description of Indians standing on their horses' backs that closes "The Desert Pampas" the image of the past is preserved as if photographed in Hudson's mind but nonetheless remains (he insists while representing it) unrepresentable. There, as well as here in this final chapter of the book, the vibrant quality of Hudson's memories, whether of pampas grass, briefly glimpsed birds, or "savages," functions to deepen the grief at the loss of what is remembered. We might conclude that nearly as threatening to Hudson as loss itself is the memory of loss. The final chapter of *The Naturalist in La Plata* exemplifies the shape taken by that threat in that it repeats, as if in miniature, the form of the entire book: a litany of remembered disappearances, things seen and lost. Such repetition constitutes not simply a reprise but an intensification of the disaster at the book's center. Multiplying and accumulating the instances in which only recollection remains to attest to the remarkable or beautiful once witnessed, it also reveals the suffering that derives from memory.

This brings us to the chapter's and the book's final pages, which chronicle the experience of seeing and losing a "savage." Stopping for a drink in a *pulperia* or roadside bar, Hudson is confronted by a man with "black eyes … even more like those of a rapacious animal in expression than in the pure-blooded Indian" and a mouth "twice the size of an average mouth" filled with teeth like those "of a shark, or crocodile" (*NLP* 376). These physical features, bizarre enough in themselves, are accompanied by an even more affecting "mental strangeness, showing itself at unexpected times, and which might flash out at any moment" (*NLP* 377). The immediate problem for Hudson is one of explanation: "[H]ow, in the name of Evolution, did he come by [those teeth], and by other … peculiarities … which made him a being different from others – one separate and far apart?" (*NLP* 378). Like the question itself, the alternatives presented by way of answer derive from the certainty that biology, and specifically evolutionary biology, holds the key:

Was he, so admirably formed, so complete and well-balanced, merely a freak of nature, to use an old-fashioned phrase – a sport, or spontaneous individual variation – an experiment for a new human type, imagined by Nature in some past period, inconceivably long ago, but which she had only now, too late, found time to carry out? Or rather, was he like that little hairy maiden exhibited not long ago in London, a reproduction of the past, the mystery called reversion – a something in the life of a species like memory in the life of the individual, the memory which suddenly brings back to the old man's mind the image of his childhood? (*NLP* 379)[35]

Remarkable about these possibilities is how similar, at first, they appear to be. The initial answer, posed in the form of a question, explains this strange personage as a *lusus naturae* or "sport," but a belated one; the alternative answer, again cast as a question, imagines him as an atavism.[36] In both instances something properly belonging to the past has appeared in the present. But the second explanation differs from the first in viewing this "singular variation in the human species" not as a merely individual variation but rather as a common ancestor to modern humans, "a reproduction of the past" (*NLP* 379). Although Hudson refuses to choose between these possibilities – I will suggest why in a moment – he displays more interest in the second, continuing to work out its entailments by noting that because "old Palaeolithic man," although possessed of "an unspeakably savage and … repulsive and horrible aspect," had teeth "not unlike our own," the man in the *pulperia*, if an atavism at all, must hearken back to "a remoter past, a more primitive man, the volume of whose history is missing from the geological record" (*NLP* 379). Alluding to Charles Lyell's and Darwin's trope of the geological record as a vast library with more missing than extant volumes, Hudson locates the temporal origin of his shark-toothed throwback in an abyssal time, a past so distant as to have left no fossil traces.[37]

Despite positing such an explanation, however, Hudson does not endorse it. For the rationale for including this description of an encounter with savagery in "Seen and Lost" is that it, too, constitutes a case of a rare individual glimpsed but not collected and thus neither documented nor classified. Hudson opens the paragraph quoted from above with the following admission:

I have never been worried with the wish or ambition to be a head-hunter in the Dyak sense, but on this one occasion I did wish that it had been possible, without violating any law, or doing anything to a fellow-creature which I should not like done to myself, to have obtained possession of this man's head, with its set of unique and terrible teeth. (*NLP* 378)

Although hedged round with qualifications necessitated by the difference made by the fact that, however savage, the head in question is nonetheless human, Hudson's desire to possess it partakes of scientific tradition – specifically, the "international trade in body parts" that Tim Fulford, Debbie Lee, and Peter J. Kitson show accompanied and enabled the elaboration of a science of race in the eighteenth and nineteenth centuries.[38] The mention of head-hunting in "the Dyak sense" seeks to distinguish it from Hudson's desire to commit the same act in what might, by contrast, be called the scientific sense, as do the final few sentences of the chapter, which turn on a fantasy of the taxonomic confusion and disagreement that would result were Hudson to "drop it [the head] like a new apple of discord, suited to the spirit of the times, among the anthropologists and evolutionists generally, of this old and learned world" (*NLP* 380).

Such an aspiration, Hudson implies, is not "savage" at all insofar as it may be included in what are for biologists among the highest ambitions imaginable, those of working out the relations among the various classes of living beings. But this implication is undercut both by the violence at its base – the severing of a head admittedly human, whatever its teeth might look like – and by Hudson's avowal that at issue for him is not taxonomy itself but rather "the pleasure – not a very noble kind of pleasure, I allow – of witnessing from a safe hiding-place the stupendous strife that would have ensued" had he brought this head to the attention of Europe's scientists (*NLP* 380). The head-hunting would be "savage," but so, too, would the sort of joy experienced at witnessing the bitter disagreement among scientists. Which is just to say that, for the first time in his natural history writing, it is as if contact with a "savage" animates the savage within.[39] Because, looking back at the options given in explanation of the man with the teeth, it is significant that the second, reversion, is called "a something in the life of a species like memory in the life of the individual, the memory which suddenly brings back to the old man's mind the image of his childhood." Although he had never before "been worried with the wish or ambition to be a head-hunter in the Dyak sense," in the presence of this readily but divergently explicable instance of savagery Hudson finds himself in danger of reverting to his own savage "childhood."

The Naturalist in La Plata begins and ends with memories of savages: at the outset, the frontispiece of Native Americans on horseback that is also a representation of Hudson's recollection of his last sight of the pampas at its most beautiful; at the conclusion, the recollection of a shark- or crocodile-toothed man who himself recalls a vision of humans that existed in an unrecorded abyss of time preceding the Paleolithic. The relation between these bookends and what comes between them – the detailed evocation of and mourning for a natural world seen and lost – is vexed. In one sense, all are to be taken together as that which has disappeared in the wake of the settlement and enclosure of the pampas. Nature and savages for Hudson are thus interimplicated with and in some ways identical to one another. But in another sense they are distinct from or even opposed to one another. Armed by the so-called civilized, for instance, modern-day savages participate in the destruction of nature – a role reprised in *Green Mansions*, when the horrible Cla-Cla and her tribe hunt down and murder Rima, apotheosis of the supernatural in the natural, "bright being, like no other in its divine brightness, so long in the making."[40] Only at the very end of *The Naturalist*, with the appearance of a savage sport or atavism and Hudson's yearning to "collect" his head, does Hudson hint at a sense of himself as savage and his understanding of memory as the faculty that can return him and other "civilized" people to a savage state. The development of this line of thought, however, would not have to wait long – only until his next collection of essays on South American natural history.

2

That collection is *Idle Days in Patagonia*, published in 1893. In its final two chapters, "The Plains of Patagonia" and "The Scent of an Evening Primrose," Hudson elaborates a theory of time, memory, and the senses that posits savagery as a forgotten but imperishable inheritance capable, in the presence of certain landscapes or smells, of being recalled and, hence, restored. In order fully to understand those chapters, it will be necessary first to consider the earlier parts of the book, for there Hudson develops his account of South America as a *lieu de mémoire* or "site of memory" and, in so doing, firmly establishes his sense of the human temporalities implied by evolutionary theory.[41]

Idle Days begins as a reprise of Kingsley's *At Last; A Christmas in the West Indies*. Titling the first chapter of his book "At Last, Patagonia!," Hudson frames it, like Kingsley frames his own travel narrative, as the chronicle of a long-awaited journey to a fabled land. This gesture immediately sets *Idle Days* apart from *The Naturalist in La Plata*. The latter purports to give a native's account of the pampas, whereas the former approaches Patagonia as a strange and little-known destination, as exotic to the author as to most of his readers. Thus, for instance, Hudson attempts to convey his reaction to seeing snow for the first time in his life by likening it to Kingsley's experience when first encountering the tropics: "only those of my English readers who, like Kingsley, have longed for a sight of tropical vegetation and scenery, and have *at last* had their longing gratified, can appreciate my sensations on first beholding snow."[42] Precisely because of its emphasis on exoticism and novelty, however, Hudson's book, if in some ways paralleling the trajectory of Kingsley's, in another way inverts it. For Kingsley, although he had never been there before, viewed the West Indian tropics and the rest of the Americas for which they stand as his rightful home; further, because he understands his presence there as both analeptic and proleptic, that presence is necessarily spectral.[43] Hudson, by contrast, finds himself the only tangible entity in a land thronged with ghosts.

A vast thanatopia, Patagonia functions for Hudson as a *lieu de mémoire* in the sense I am using the term in this book: a place that by virtue of its own perceived archaism returns those who traverse it to the past. Hudson hints at this function for the South American continent as a whole in the first chapter of *The Naturalist in La Plata*, where he writes: "I love the living that are above the earth; and how small a remnant they are in South America we know, and now yearly becoming more precious as it dwindles away" (*NLP* 15–16). In keeping with the overriding concerns of that book, this statement relates most directly to Hudson's catalogues of destruction and extinction. The emphasis falls not on the catacomb full to overflowing with the bones of the dead but on the continued disappearance of the "living remnant" still "above the earth." Viewed in light of *Idle Days*, however, the import of the sentence shifts, for in the later book the vision of Patagonia as an immense tomb is both paramount and constitutive of its peculiar promise.

In the third chapter of *Idle Days*, "Valley of the Black River," Hudson describes the area surrounding the Río Negro as a literal graveyard scattered with the shattered skulls of peoples who lived there before European colonization (*IDP* 38–39). The point is to distinguish not only past from present but also America from Europe, barbarism from civilization.[44] But when he returns to these skulls in the next chapter, "Aspects of the Valley," he does so to advance a view of them not as reifying boundaries between that set of binaries but rather as allowing, in fact requiring, their dissolution:

If by looking into the empty cavity of one of those broken unburied skulls I had been able to see, as in a magic glass, an image of the world as it once existed in the living brain, what should I have seen? Such a question would not and could not, I imagine, be suggested by the sight of a bleached broken human skull in any other region; but in Patagonia it does not seem grotesque, nor merely idle, nor quite fanciful … On the contrary, it strikes one there as natural; and the answer to it is easy, and only one answer is possible. (*IDP* 40)

Although (and also because) the hole in a "bleached broken human skull" contains nothing, it serves as a window that reveals the lineaments of an ancient landscape as it appeared to one of its human or proto-human inhabitants. Or, I should say, it *might* reveal such things – for Hudson casts this proposition in the conditional: if he could see a vanished world in this skull, he wonders, the world as it appeared to the once-living human who looked out of eye sockets that are now, like the skull itself, empty, what would that world look like? Certainty is immediately reasserted in the sentences that follow, which answer both parts of this question by declaring the first, with its idea of a broken skull as a "magic glass" capable of providing a glimpse into the past, "natural" in Patagonia, and the second, with its query about what that glass would show, "easy" to answer because "only one answer is possible."

The answer proves singular and simple because the person of which the skull is a remnant would have seen precisely what Hudson sees: the Río Negro itself, which Hudson argues "must have been to the inhabitants of the valley the one great central unforgettable fact" (*IDP* 41). But Hudson's ability to divine what counted as supremely important to those aboriginal inhabitants derives as well from his powerful identification with them, an identification also signaled by his description of the hole in the skull as a "magic glass." If that glass is in one sense a window, in another it is a mirror. And since, further, a mirror gives back a reflection of the person looking into it, the "image of the world as it once existed in the living brain" must be an image of Hudson himself.

At the end of the *Descent*, Darwin writes: "The astonishment which I felt upon first seeing a party of Fuegians on a wild and broken shore will never be forgotten by me, for the reflection at once rushed into my mind – such were our ancestors."[45] Fuegians act as a mirror for Darwin in that they give rise to a reflection: a thought and a recollection but also a picture of his progenitors. But if past and present collapse into one another for Darwin, actually existing peoples incarnating long dead ancestors, they are also kept apart. For, except very clandestinely, he does not see in Fuegians an accurate reflection of himself. In *Idle Days in Patagonia*, by contrast, past and present become so entangled as to be nearly indistinguishable. The magic glass of a smashed Patagonian skull reflects an image of the past that is also an image of the present, a vision of a prehistoric world as seen by one of its inhabitants that is also a vision of Hudson. Savage and civilized, ancient and modern run together, fused in the person of Hudson, who is now both the civilized observer and the savage observed, both a modern student of the distant past and an avatar of that past.

What makes such multiplicity possible, as Hudson writes elsewhere in *Idle Days*, is his conviction that not he alone but all "modern" and "civilized" people are, at base, savages: "in our inmost natures, our deepest feelings, we are still one with the savage" (*IDP* 214). But if that is so, why do some places lend themselves to the re-emergence of this inner savage more than others? Why, for instance, does the question of what vision of the past is revealed in the cavity of a broken human skull, which "in any other region" would seem "grotesque," "idle," or "fanciful," seem not only reasonable but easily answered in Patagonia? To address these questions is to go to the heart of Hudson's peculiar collocation of memory, evolutionary theory, and human being. And it is again the third chapter of *Idle Days*, "Valley of the Black River," that provides one of the keys to the whole. There, writing of "efforts to know more, or to imagine more" about mind of the "savage," Hudson declares: "By taking thought I am convinced that we can make no progress in this direction, simply because we cannot voluntarily escape from our own personality, our environment, our outlook on nature." In a relativism that is at the same time a determinism, Hudson posits the existence of a savage and a civilized self, and savage and civilized relations with the natural world, even as he seems to close down the possibility of moving between them. As he immediately makes clear, however, the inability "voluntarily" to leave ourselves behind for other, earlier versions of self does not mean there is no hope of getting free. We can escape, but only, as it were, by accident – only when such escape is "unsought" and unexpected (*IDP* 37–38).

It is of course memory for Hudson that allows an escape from the modern, civilized self, a return to savagery. But this passage in *Idle Days* signals that Hudson recognizes more than one kind of memory. As for Proust, so for Hudson memory falls into distinct categories, voluntary and involuntary. Moreover, not only do both writers assert this distinction, both also valorize involuntary memory for its transformative powers. In the "Overture" to *Swann's Way* (1913), the first volume of *The Remembrance of Things Past*, Proust's narrator famously avers:

And so it is with our own past. It is a labour in vain to attempt to recapture it: all the efforts of our intellect must prove futile. The past is hidden somewhere outside the realm, beyond the reach of the intellect, in some material object (in the sensation which that material object will give us) which we do not suspect. And as for that object, it depends on chance whether we may come upon it or not before we ourselves die.[46]

The deliberate, intellectual effort to recall the past cannot succeed. It is possible to think back to a particular moment in one's life and remember certain of its aspects; Proust's narrator is not denying the existence of memory in this sense. But such a memory proves worthless for the enterprise the narrator engages in, an enterprise whose full reach is conveyed in the word "recapture."[47] If one wishes to recapture the past – to "give [it] reality and substance," as the narrator writes elsewhere in the "Overture" – it is not to voluntary or intellectual memory that one must turn.[48] For access to the past, in this account, resides not in ourselves, and thus nowhere that intellectual memory is capable of reaching, but in "some material object" out in the world, and in "the sensation which that material object will give us."

Several far-reaching entailments follow from these claims. To begin with, as the narrator points out, the past might well be lost forever – might never be recaptured or given substance again. We could ourselves die before resurrecting the dead past, even a past so rich and apparently unforgettable as that given novelistic form by Proust. Moreover, whether the past lives again or not is, purely and simply, a matter of chance. We may encounter the fateful object that will open the door to the past; we may not. The entire enterprise is fragile, shadowed by the possibility, even the likelihood, of failure. Finally – something the narrator does not state directly but that is everywhere implied by the astonishingly loving and detailed attention to (to mention a few examples from hundreds) steeples, trees, paths, and cookies in *The Remembrance of Things Past* – the past, and hence memory, commonsensically imagined to belong to persons, actually belongs to things. Because memory resides in things, not the inner mental theater of the subject, the whole of the object world is luminous with potential. It happens that the taste of a madeleine soaked in linden-flower tea gives rise to the narrator's memories of his childhood at Combray, but it need not have been that way. Any object at all may have accomplished the same thing, any taste or smell produced by that object, anywhere. Proust thus anticipates Pierre Nora, who writes: "Since no one knows what the past will be made of next, anxiety turns everything into a trace, a possible indication, a hint of history that contaminates the innocence of all things."[49] It is as though the "Overture" to *Swann's Way* transmogrifies the material world in its entirety into a sea of faces that may contain, somewhere in its midst, the visage of a friend one thought had died.

The metaphor is not mine but Hudson's, who twice deploys it in *Idle Days* to describe the sensation of the unexpected advent of a once-forgotten memory (*IDP* 49, 232). On the strength of this and other parallels between *Idle Days* and *Swann's Way*, we could almost say that Hudson's understanding of memory is Proustian *avant la lettre*. But Hudson's evolutionism makes for crucial differences. For him not only the personal past but also the species past – not only, to invoke the terms of Ernst Haeckel's biogenetic law, ontogeny but also phylogeny – can be recaptured via memories triggered by an encounter with some aspect of the material world. Further, if those memories are in Hudson's account, too, properly involuntary – if there is no possibility of simply willing oneself to recollect what it was like growing up on the pampas – it is nevertheless at least possible to place oneself in a setting where that recollection's occurrence will be more probable.[50] The absolute contingency that Proust imagines to appertain to the project of recapturing the past can, for Hudson, to some extent be mitigated. Certain places, in particular, hold more promise than others for the return to childhood – one's own or that of one's species.

Patagonia is such a place. Even if the memory of being savage cannot deliberately be recalled, its surfacing need not be left entirely to chance. We can, Hudson suggests, court that memory by placing ourselves in a position to receive it. Itself understood to be an anachronism, the repository of the storied dead and thus closer to them and closer to the past than are other areas of the globe, Patagonia serves as a site favorable to the return of memories, and in particular to the memory of having been savage. This aspect of Hudson's thought was not lost on his contemporaries. Wallace addresses it directly when, reviewing *Idle Days* in *Nature*, he comments: "There was nothing beautiful or even pleasing to be seen in this dreary monotonous solitude, yet he [Hudson] felt a great delight and satisfaction in it, which he imputes to the ancestral savage nature that still exists in all of us, though repressed and overlaid by civilisation and society."[51] If Wallace omits to mention the complex mechanism that connects Patagonian monotony to the pleasure Hudson finds by way of the relay of involuntary memory, he nonetheless conveys the essential prerequisite to its workings: the conviction that an "ancestral savage nature … still exists in all of us."

Hudson develops this line of thought most fully in the last two chapters of *Idle Days*. Chapter 13, "The Plains of Patagonia," seeks to establish the validity of the claim Wallace repeats: that, as Hudson has it on the last page of that chapter, "in our inmost natures, our deepest feelings, we are still one with the savage" (*IDP* 214). Hudson offers proof of that claim in the form of his own experiences of returning to a savage state, which underwrite the validity of his understanding of the continued existence and availability of the deep past, specifically the evolutionary past, within each organism. Summing up that understanding, he crafts a description of modern humans as themselves *lieux de mémoire* – catacombs filled with the remains of a world that has perished but that is subject to resurrection by means of recollection: "And we ourselves are the living sepulchres of a dead past – that past which was ours for so many thousands of years before this life of the present began; its old bones are slumbering in us – dead, and yet not dead nor deaf to Nature's voices" (*IDP* 206).[52]

The slippage between place and person is instructive. We have seen it before when, in "Aspects of the Valley," Hudson wonders what he would see if, looking into the hole in a broken skull, he could glimpse "an image of the world as it once existed in the living brain" – and sees an image of himself. But what might seem incoherent or rhetorical sleight of hand turns out to be a constitutive feature of Hudson's writing and thinking, and one that further points his difference from Proust. For if, in *Swann's Way*, subject and object are distinct, and it is in the object world that access to the past of the subject resides, in *Idle Days* subject and object, modern human and South American landscape, mirror one another. Both exist in the now, the present moment, but both contain in themselves, buried but capable of being disinterred, the remains of a past that made them what they are. They share a savage past that each may summon in the other, a past that belongs properly to neither but returns or can be recalled when the two come into proximity.

Such a distinction inheres in the fact that for Hudson the human subject is first and foremost the subject of evolution. And if "The Plains of Patagonia" establishes that fact, the final chapter of *Idle Days*, "The Scent of an Evening Primrose," constitutes something like a case study of its consequences. Rehearsing and reinflecting his theory of memory such that the personal rather than species past comes to the fore, Hudson at first seems to leave behind questions of evolution and savagery altogether. But the exploration of sensory perception in relation to memory that he elaborates in fact derives its subtending logic from the same evolutionary paradigm that governs the rest of his thought.

The title of the chapter alludes to Hudson's habit, in an English garden, of smelling primroses, flowers familiar from his childhood and youth. Like setting off for Patagonia, such smelling is an attempt to position himself to allow the return of a memory. Apparently more reliable a mnemonic than days spent alone in the South American wilderness, the scent of the evening primrose never fails to produce the memory sought, a memory that Hudson describes as follows:

For a space of time so short that if it could be measured it would probably be found to occupy no more than a fraction of a second, I am no longer in an English garden recalling and consciously thinking about that vanished past, but during that brief moment time and space seem annihilated and the past is now. I am again on the grassy pampas, where I have been sleeping very soundly under the stars. (*IDP* 222)

The recollection that results from sniffing a primrose is neither a reminiscence about the past nor the past's recurrence in the present. It is, rather, a return: "[T]he past is now. I am again on the grassy pampas." As Hudson puts it at a later moment in the chapter, the "suddenly recovered sensation is more to us for a moment than a mere sensation; it is like a recovery of the irrecoverable past" (*IDP* 232).

If this amounts to a theory of memory – memory as more than notional, memory as time travel – it is also, just as crucially, a theory of the senses. For the place of smell in this account is far from incidental. On the contrary, only smell is capable of acting in this way. "Nothing that we see or hear," declares Hudson, "can thus restore the past" (*IDP* 223). This is because, unlike sights and sounds, smells "cannot … be reproduced in the mind" – cannot, that is, deliberately be recollected (*IDP* 226). It is as if the workings of involuntary memory were built into the sense of smell – as if in its very essence it partook of that Proustian dictum about the contingency of the return of the past. But, says Hudson, it wasn't always so: he postulates a former ability to recall smells, an ability retained by "some savages" because smell is more important to them than it is to the "civilized" (*IDP* 227, 228). Here is the beginning of an evolutionary history of sensory perception, a project to which Hudson would return in the last book he wrote, *A Hind in Richmond Park*. In that book, incomplete at the time of his death, he gives final form to his hopes for a return to savagery as recompense for and counter to the loss of wild nature.

3

Long before Diane Ackerman's *A Natural History of the Senses* (1990) or Michael Taussig's *Mimesis and Alterity: A Particular History of the Senses* (1993), Hudson produced in *A Hind in Richmond Park* what both Ackerman's and Taussig's titles purport to offer: an investigation of human sense perception understood as at once natural, part of *Homo sapiens'* biological inheritance, and historical, having undergone change on a human time-scale rather than a geological one.[53] Reading it therefore involves particular attention to the historicity of the senses. But such reading also demands awareness of what might be called the historicity of the historicity of the senses: the various ways that such a history has been or might be told. Hudson's book is distinctive in this regard both for its contention that humans possess more senses than what he calls "the canonical five" as well as for its thoroughgoing evolutionary perspective, which finds adaptation to changed environment and mode of life to be the key to changes in the human sensorium (*HRP* 18).[54]

Among those senses we possess that have heretofore gone unstudied, Hudson contends, are the "atmospheric sense," or responsiveness to wind; the "sense of direction"; the "migratory sense," which Hudson insists is distinct from the sense of direction; and the "sense of polarity," the ability to perceive the body's alignment in relation to a north-south axis (*HRP* 18, 141, 211). Apart from a few chapters devoted to bird migration that seek to explain this phenomenon by appeal to that heretofore unknown migratory sense, however, and some thoughts on the connection between the atmospheric sense and telepathy, the bulk of the book actually focuses on two familiar perceptual capacities rather than any Hudson purports to have discovered. Most at issue in his treatment of these decidedly canonical senses, hearing and smell, is the extent to which they have been subject to alteration over time: the diminution of the acuity of the ability to smell from animals to prehistoric humans and then again from prehistoric to modern humans, for instance. In addition to documenting such change, Hudson speculates on the nature and causes of the effects of stimulating these senses. He attempts to specify what one feels when listening to music, and why one feels it; or, to take another example, he suggests reasons for the intimate connection between scents and memories.

The place of evolution in the book may be established by appeal to its first few pages. In an anecdote that both explains his title and lays the groundwork for the treatise that follows, Hudson recounts a visit to London's Richmond Park during which he finds himself closely observing the rapidly moving ears of a hind or female deer. The hind engaged in "a constant succession of small movements … [that] told their tale – a sudden suspension of the cud-chewing, a stiffening of the forward-pointing ears, or a slight change in their direction"; "she was experiencing a continual succession of little thrills," he concludes, and attributes the source of those thrills to sounds she was able to hear but he was not (*HRP* 4). Hudson insists that these movements, because not simply instinctual but in fact "saturated with intelligence," demonstrate that the "mind-life" of animals is fundamentally "similar to, though much lower in degree than, that of man" (*HRP* 6, 7).[55] He further claims that animals therefore provide clues as to the mental and physical traits of ancient as well as contemporary humans. Extrapolating from what he has seen with that claim in mind, Hudson immediately goes on to argue that, like the ears in deer, the "ears in primitive man were free-hinged, not nailed to the head." They

were also undoubtedly very much bigger than ours ... I dare say that the Palaeolithic man's auricle was about the bigness of a tea-cup half-saucer, and being hinged, it could lie back flat against his head when listening to sounds from the sides or the rear ... [as well as] stand out, as in the elephant, at right angles to the head. Judging from the position of the ears in new-born babies, one might suppose that the ears ordinarily stood out from the head like the two opposite handles of a round pot. (*HRP* 12–13)

This initial comparison of the hind's and Paleolithic humans' mobile ears opens out into the distinct lines of inquiry that comprise the remainder of *A Hind*. In one, which continues to develop the implications of the continuity among nonhuman animals, proto-hominids, and modern humans, Hudson discusses those primitive sensory capacities that moderns have lost, retained in a diminished form, or, in cases of atavism, suddenly regained. In that discussion's most astonishing moment, Hudson describes the unexpected return of his own "sense of direction," writing:

it affected me like the recovery of something infinitely precious, so long lost that I had been without hope of ever finding it again; and it was like the recovery of sight to a blind man; or like that "vision of Paradise" which a temporary recovery of the sense of smell had seemed to Wordsworth as he sat in a garden full of flowers; or like the recovery of memory in one who had lost that faculty. (*HRP* 148)

As such a passage indicates, this line of argument reveals Hudson's longing for the past to be a kind of primitivism of perception. The lost world to which he would return promises no utopian social organization, no freedom from the constraining sexual mores of "civilization," no prospect of a life free from the necessity of labor, nor indeed any of the usual desiderata of nostalgic utopias. His vision of Paradise lost is, rather, that of a sensorial Eden – of a time when the senses, canonical and apocryphal, possessed to the fullest powers now attenuated or wholly absent.

It is thus to be expected that, in the book's other line of argument, which takes up the question of the affective response to particular sensory stimuli, Hudson finds himself itemizing those stimuli that are in the process of disappearing or that have already disappeared. At the conclusion of a lengthy meditation on the "beautiful wild trisyllabic alarm cry of the upland plover" that haunted his childhood, for example, he writes:

"This sound lives in memory still, but is heard no more, or will shortly be heard no more, on earth, since this bird too is now on the list of the 'next candidates for extinction'" (*HRP* 171, 173). Such a matter-of-fact observation metamorphoses, in the paragraph that follows, into a lyrically plangent depiction of worldwide ecological devastation:

All this incalculable destruction of bird life has come about since the seventies of the last century, and is going on now despite the efforts of those who are striving, by promoting legislation and by all other possible means, to save "the remnant." But, alas! The forces of brutality, the Caliban in man, are proving too powerful; the lost species are lost for all time, and a thousand years of the strictest protection … would not restore the still existing bird life to the abundance of half a century ago. The beautiful has vanished and returns not. (*HRP* 173–74)

Highly reminiscent of similar passages in *The Naturalist in La Plata*, this grim prediction establishes Hudson's own estimate of the value of his work as a record of former perceptual wealth for a generation entering a time of scarcity.

A Hind's investigation into the historicity of the senses thus splits apart into what might usefully be called an "intrinsic" and an "extrinsic" history: a history of change in the senses of sight, smell, and hearing themselves accompanied by a history of changes in the world that alter – in fact, for Hudson, that nearly always diminish – that which exists to be seen, smelled, or heard.[56] But these two kinds of history converge, and it is that convergence that provides an overarching unity to this book, which can often seem to comprise an incoherent set of notes joined by little but the fact of having been penned by the same person.[57]

To take only one from many possible illustrative passages: in chapter 16, which considers the origins of human musicality and rejects Darwin's explanation of those origins in sexual selection, Hudson mentions in passing Ludwig von Beethoven's modest observation that "those who listened to [his music] were lifted above this earth into a higher sphere and state" (*HRP* 249). Hudson does not directly contest Beethoven's account of his power as a composer so much as set it aside in favor of an alternative one:

It may be so: I do not know; but I do know that it takes me back; that it wears an expression which startles and holds me, that it is essentially the "Passion of the Past" – not of mine only, my own little emotional experiences, but that of the race, the inherited remembrances or associations of its passionate life, back to a period so remote that it cannot be measured by years. A dreadful past, but at so great a distance that it is like the giant terrifying mountain, the heart-breaking stony wilderness with winter everlasting for its crown, seen afar off, softened and glorified with rose and purple colour, at eventide. (*HRP* 249)

Such a meditation, although framed as a characterization of the effects of listening to the work of a particular composer, constitutes a claim about the effect of music as such. Beethoven in this regard is representative rather than unique. All music, Hudson argues, must be understood chiefly in relation to its ability to spur recollection. Further, because music dates from the first moment of "the marriage of sound with emotion," encounters with music, especially powerful ones, return the listener to that time (*HRP* 249).[58] And although when approached from the present it takes on an aspect of sublimity rather than outright peril, the moment in question belongs properly to the "dreadful," "terrifying," and "heart-breaking" beginnings of "the race." It belongs, that is, to the "savage" past.

Thus it is in connection with the savage that the extrinsic and intrinsic histories of the senses that make up *A Hind in Richmond Park* turn out to be the same history. Changes in the human sensorium as well as changes in the world to be sensed, Hudson maintains, always amount to loss. Species are passing away into oblivion, and with them sights, sounds, scents; at the same time and partly as a result, human perceptual capacities, already much reduced in relation to what they once were, face continued atrophy. The intimate interdependence of human perception and that which exists to be perceived means that to destroy the latter is, inevitably, to diminish the former. By the same token, however, such interdependence also holds out the promise of a return to the lost past. The sound of Beethoven's music, the scent of an evening primrose, the landscape of Patagonia: all take one back. By way of the memories they evoke they effect a return, and a return not to any distant time but specifically to that of human prehistory.

Hudson's task is thus to enlist his own personal memories in the service of summoning a collective memory of what humans once were. For if the recollection of the cry of the upland plover or of "Indians on the Lookout for Strayed Horses" is in some sense Hudson's alone, it is through them that he seeks to convey another recollection, that of the "dreadful past" that belongs to us all. *The Naturalist in La Plata*, *Idle Days in Patagonia*, *A Hind in Richmond Park* – these books are designed to function both as instigations to remembering and prosthetic memories in themselves. An appreciation of Hudson that appeared in *The Morning Post* makes the point directly: "There is a magic in Mr. Hudson's style and in his exquisite sensibility which awakens in his reader a thousand sleeping memories."[59] Like Beethoven, a primrose, or Patagonia, Hudson's works of natural history remind readers of what they might not have known they had forgotten: that they are savages – and that, as such, to destroy wildness is to destroy themselves.

4

In a tour de force of popular evolutionary paleontology titled *Wonderful Life: The Burgess Shale and the Nature of History* (1989), Stephen Jay Gould, Hudson-like, narrates the history of life on the planet as a history of loss. The book details how, in 1909, an American geologist named Charles Doolittle Walcott discovered a fossil bed in British Columbia that would come to be known as the Burgess Shale; how Walcott identified all the bizarre soft-bodied organisms he found there, organisms dating to just after the Cambrian explosion of 570 million years ago, as primitive precursors of modern organisms; and how, in the 1970s and 1980s, a group of researchers based at Cambridge began re-describing the fossils, eventually concluding that most of them cannot be directly allied with modern organisms at all and that, further, "the creatures from this single quarry in British Columbia probably exceed, in anatomical range, the entire spectrum of invertebrate life in today's oceans. Some fifteen to twenty Burgess species ... should probably be classified as separate phyla."[60]

However remarkable to specialists in invertebrate paleontology, that conclusion may not surprise the lay reader. But in Gould's view it amounts to nothing less than a reversal of what had previously been assumed about life on earth. The Burgess Shale organisms demonstrate, he claims, that the history of that life does not reveal ever-increasing diversity but rather an early efflorescence of diversity followed by mass extinction and then diversification within a few remaining basic anatomical plans (*WL* 46).[61] Moreover, given what he understands as the wholly contingent process that resulted in the disappearance of most of those plans and the persistence of a handful, Gould argues that the Burgess Shale militates against the applicability of Spencer's "survival of the fittest" at this early moment, suggesting instead "random elimination of most lineages" (*WL* 47). The creatures that survived did so, Gould avers in what is for a self-avowedly passionate evolutionist a shockingly un-Darwinian claim, not because they were somehow better adapted to changing conditions but because they were lucky.

In the wake of the re-description of the Burgess Shale organisms we must no longer envision the history of life as a gradual, inevitable, predictable movement; we must instead, Gould reasons, see it as "a staggeringly improbable series of events, sensible enough in retrospect and subject to a rigorous explanation, but utterly unpredictable and quite unrepeatable" (*WL* 14). This change in turn forces a re-imagining of what might well be named human ethics. Because humans have been produced by that very improbable, unpredictable, and unrepeatable process, we are contingent creatures rather than the telos of some biological tendency. Like the surviving Burgess Shale organisms, we have simply been very lucky. "We are the offspring of history," Gould writes in his final sentence, "and must establish our own paths in this most diverse and interesting of conceivable universes – one indifferent to our suffering, and therefore offering us maximal freedom to thrive, or to fail, in our own chosen way" (*WL* 323).

I find *Wonderful Life* Gould's most moving book. From unpromising materials – the reconstructed remains of soft-bodied organisms that lived in the ocean nearly 600 million years ago – he rewrites the story of life, locating its heyday in a distant past now lost forever. He evokes a Cambrian world about which I find I care and for whose loss I almost mourn. Even his vision of a universe "indifferent to our suffering, and therefore offering us maximal freedom," despite bearing a close resemblance to mid-twentieth-century existentialist philosophy, seems more persuasive than that philosophy insofar as it has been arrived at by means of a close analysis of the fossil record. But just here one encounters a disturbing paradox. For *Wonderful Life*, the title of which attests to a celebratory attitude to living beings *in toto* (and now-extinct invertebrates in particular), is nonetheless informed by an overarching anthropocentrism. Despite the carefully crafted accounts of the creatures of the Burgess Shale and the resulting sense of loss at the disappearance of most of them together with the phyla to which they belonged, by the end of the book it seems they have not been the point at all. Everything redounds to humans – to how we should live, to what we ought to think of the world and our place in it. Not only does this seem strange in itself, given the context, but it also somehow undercuts the force of the argument about extant species as, finally, random products of the evolutionary process. It is as if Gould were turning around on himself and, having dismissed the notion that we are chosen beings by demonstrating that we are simply lucky, concludes suddenly that because we are lucky we must see ourselves as chosen.

Of course it is entirely "natural" for us to be most interested in ourselves. But some humans manage not to be, and one of the most remarkable and instructive aspects of Hudson's work is how he turns again and again to nonhuman animals when marveling at beauty and mourning its loss. Describing the charge of the evolutionary biologist, Gould writes: "[W]e are trying to account for uniqueness of detail that cannot, both by laws of probability and time's arrow of irreversibility, occur together again" (*WL* 278). Precisely such uniqueness points Hudson's agony at widespread extinction. Gould inherits and develops Hudson's sense of evolution as a theory that makes the life of each species precious by making it unlikely and, more significantly, unrepeatable. But Hudson, unlike Gould, does not limit the resulting ethical stance to the human. All species are unlikely, all may appear once and once only, and all must be protected when possible and commemorated when not.

But there is in Hudson, too, a kind of human exceptionalism, and it has to do with the savage. Crucially, it also has to do with the kind of memory associated with a return to savagery. Recalling lost natural beauty and lamenting its disappearance involves Hudson in a controlled, deliberate mnemonic practice. The claim to a literal return to savagery, on the other hand, relies on the workings of a memory that is corporeal and involuntary. "By taking thought," writes Hudson about the attempt to fathom the workings of the proto-human or savage mind, "I am convinced that we can make no progress in this direction, simply because we cannot voluntarily escape from our own personality, our environment, our outlook on nature" (*IDP* 38). In this regard Hudson's "taking thought" suffers from the same failing as Marcel Proust's *mémoire volontaire*. Walter Benjamin's gloss on the latter well explains the problem with both: "the information which it gives about the past retains no trace of it."[62] Deliberation is powerless to resurrect the past within. Nevertheless, Hudson insists, such resurrection is possible – but only by way of sensory experience and the recollection it occasions. Certain landscapes, smells, and sounds, encountered unawares, can literally take one back. The personal or private past of the individual may be returned to in this way, but the more consequential return such desultory or involuntary memory enables is that of the species past.[63] The descendant of savages, each modern human contains his or her progenitors within, and a specific kind of memory may, under certain conditions, disinter what usually remains buried. But the possibility of this access to the savage past is rendered tenuous because it cannot be had on purpose. Contingent, aleatory, it is without guarantee and impervious to the promptings of desire.

This points the difference between other kinds of memory and the memory of the savage, explains Hudson's human exceptionalism, and also explains the privileged place savagery holds in his imagination. His identification with the savage carries with it the same enormous political and ideological risks of all forms of primitivism – dangers anatomized in the work of scholars ranging from Peter Mason and Edward Said to Anne McClintock and Kobena Mercer.[64] This body of work compels attention to the complicity of Hudson's thought with exoticist and fetishizing discourses that functioned (and continue to function) to deny political agency to indigenous peoples struggling against imperialism (formal or informal) and colonialism (paleo- or neo-). That complicity must always, I think, be borne in mind, in just the way that Darwin exhorts us always to remember loss and death.

But it is just here, in connection with the memory of loss and memory as loss, that Hudson marks and makes a difference. For if, in one sense, turning within and to the past in the face of present-day loss constitutes an admission of defeat, in another sense that turn also suggests that the past that exists within may be enlisted in the service of the creation of a public memory tied to political goals. It is none other than Jorge Luis Borges who best illuminates this aspect of Hudson when he writes in an appreciation of *The Purple Land*: "Macaulay, in the article about Bunyan, marveled that one man's imaginings would become, years later, the personal memories of many other men. Hudson's imaginings remain in the memory."[65] And so while at the beginning of this chapter I distinguished Hudson's activism from his work as a memorian, it is now possible to recognize that the two are of a piece – and, more specifically, that to counter the memory of loss represented by extinct species as well as to forestall the continuation of that loss into the future, Hudson crafted his natural history writings to serve as prosthetic, collective memories – reminders of a wild and "savage" past most of his audience had never experienced but that they could, in reading his books, remember.

Coda: some reflections

One of the most striking contemporary caricatures of Charles Darwin was published in *Figaro's London Sketch Book of Celebrities* on February 18, 1874 (see frontispiece). It depicts a simianized Darwin kneeling on a rough shelf of rock. Beside him sits an anthropomorphized monkey, legs crossed, left arm extended down, right hand lifted in a gesture of self-defense or possibly astonishment. That the monkey's hair seems to be standing on end militates in favor of the second alternative. Darwin looks meaningfully in the direction of the monkey. The monkey does not return Darwin's gaze but stares, instead, into a hand-mirror Darwin holds up to it. In the background, faintly rendered palm trees soar over the heads of both figures. Below the image the caption "Prof. Darwin" appears, and below the caption two quotations from Shakespeare: "This is the ape of form," from *Love's Labour's Lost*, and "Some four or five descents since," from *All's Well that Ends Well*.

Much of this image and its accompanying text make immediate sense. Like many caricatures that appeared after the publication of *The Descent of Man, and Selection in Relation to Sex* (1871), "Prof. Darwin" plays on public familiarity with Darwin's appearance combined with a literalized and simplified understanding of the case for human descent from other animals. The very recognizable biologist is subjected to the implications of his own theory about the ape-like progenitors of humanity by being transformed into a monkey himself. The two snippets of Shakespearean dialogue hammer home the point. The first puns on "ape," drawing on its substantive meaning, "a monkey," as well as on two of its verbal ones, "to imitate" and "to mock," in order to imply that Darwin's claim about human origins renders the human a mockery and thus deserves to be mocked. The second, directly invoking the title of *The Descent of Man*, recalls the book's key contention but grotesquely collapses its time frame to effect another moment of apparent absurdity: the suggestion that the simian ancestors of *Homo sapiens* existed at a distance of a mere four or five generations from late-nineteenth-century humans – nearer in time to the Victorians, that is, than Shakespeare himself.[1] Even the plays' titles contribute to the comic effect: "love's labor lost" may be a sardonic comment on the disappointing outcome of the sexual selection given so prominent a role in the *Descent*, while "all's well that ends well" signifies, in a countervailing direction, the benignity of the process that produced humans, whatever its incongruity or distastefulness.

So much seems clear. Less so is what we are to make of the monkey and the mirror – specifically, what we should imagine the monkey sees in the mirror. Janet Browne, in her comprehensive treatment of Victorian caricatures of Darwin, claims that in the *London Sketch Book* image Darwin is represented "politely inviting [the] ape to contemplate its future."[2] This interpretation has much to recommend it. Common misunderstandings and distortions of the assertion that, as Darwin writes in *The Descent*, "man is descended from a hairy quadruped, furnished with a tail and pointed ears, probably arboreal in its habits" frequently amounted to the notion of human descent from actually existing apes.[3] Modern apes are taken to represent what humans once were, modern humans what those apes have become. Browne reads "Prof. Darwin" as a visual epitome of these confused temporal relations insofar as, in claiming that Darwin asks the monkey to see itself as what it will be in the future, she seems to posit that the monkey looks not at its own image in the glass but at Darwin's. In a moment of self-reflection – something taken by many, including Darwin himself, as a hallmark of the human – the monkey finds that it is not itself but an incipient version of something else. The world evoked inside the caricature depicts a scene of pedagogy that inverts but also mirrors a scene outside it: Darwin, who taught humans to recognize themselves (their past selves or origins) in apes, teaches an ape to recognize itself (its future self or telos) in Darwin.

Read in this way, "Prof. Darwin," in spite of its evident attempt at misrepresentation, conveys a truth. Evolutionary theory destabilizes past and present, uproots identity, and puts all in motion; boundaries blur, time clots, origin and telos sit side-by-side. Reflection, in the context of evolution, reveals not the self-same but an other self, either past or to come, a not-quite or not-yet self. Pierre Nora writes: "We could speak of mirror-memory if all mirrors did not reflect the same – for it is difference that we are seeking, the ephemeral spectacle of an unrecoverable identity."[4] But Darwin holds up the very sort of magic glass Nora discounts, a mirror that gives back an image of difference rather than identity. So we can and indeed should speak of mirror-memory. That memory functions not as what holds the self together, as for instance in Frances Ferguson's explanation of Locke's account of memory "anchoring a sense of individual continuity over time," but rather as what defines the self in relation to what it once was but is no longer, what it cannot still be.[5] As the reservoir of a self that has disappeared but also oddly persists, the mirror-memory that is evolutionary theory both poses a threat and offers an ineffable promise. And because that mirror reflects precisely what Nora calls "the ephemeral spectacle of an unrecoverable identity," it mandates a memory of loss.

Evolutionary theory may therefore be viewed as both the discovery of and the response to disaster. In this it bears surprisingly close resemblance to the diverse set of present-day intellectual practices united under the umbrella of memory studies, which emerged as an attempt to document and come to terms with a series of catastrophes: the First World War; decolonization and the brutality that preceded, accompanied, or followed it; the Vietnam War; genocide; regimes kept in power by means of torture and disappearances; and, above all, the Holocaust. To speak of memory at the end of the twentieth century and the beginning of the twenty-first has been of necessity to speak of loss, and not just the memory of widespread loss but also the potential loss of memory that would expunge from the historical record the mass suffering of which so much of the last hundred years has been comprised. We could accurately say about many of the products of memory studies what Cathy Caruth says about the objects of her analysis in *Unclaimed Experience* (itself, of course, one of those products). "In these texts," she contends, "it is the inextricability of the story of one's life from the story of a death, an impossible and necessary double telling, that constitutes their historical witness."[6]

Darwin makes his argument for natural selection by reminding us of the same inextricability: "we do not see," he writes in *On the Origin of Species* (1859), "or we forget, that the birds which are idly singing round us mostly live on insects or seeds, and are thus constantly destroying life; or we forget how largely these songsters, or their eggs, or their nestlings, are destroyed by birds and beasts of prey."[7] Recalling the ubiquity of destruction stands as the *sine qua non* of understanding evolution. By the same token, evolution forces a rewriting of the history of life on the planet as a history of loss. Wallace, Kingsley, and Hudson all tell versions of that history, stories of the past destruction that gave shape to the present: the death of nearly every organism and the extinction of nearly all species and genera that lived before the moment of the writing; the disappearance of precursors and, with them, the possibility of exploration; the evanescence of "savages," contaminated by contact with "civilization" or extirpated by its representatives. These writers memorialize such disasters, but they also do more: they constitute themselves and indeed humans as such as the bearers of an almost insupportable memory.

The contemporaneity of such a constitution is indicated by its common ground with not only memory studies but also one of the major recent novelistic treatments of the dialectic of memory and forgetting, W. G. Sebald's *Austerlitz*. Sebald gives voice to a dilemma recognizable in the pages of the works of Victorian travel, natural history, evolutionary biology, and autobiography that have been my focus: on one hand, that we seem to live in a post-mnemonic age; on the other, that if, in spite of all impediments, we could somehow recall the past that increasingly eludes us, we would react with dismay. The unnamed frame narrator states the first half of this dilemma directly:

Even now, when I try to remember … the darkness does not lift but becomes yet heavier as I think how little we can hold in mind, how everything is constantly lapsing into oblivion with every extinguished life, how the world is, as it were, draining itself, in that the history of countless places and objects which themselves have no power of memory is never heard, never described or passed on.[8]

The second half is borne out by the novel as a whole, which chronicles the title character's search for a forgotten past that ends in an encounter with the dislocations and exterminations of the Holocaust. A Czech Jew sent from Prague to Wales aboard a *Kindertransport* before the outbreak of the Second World War, Austerlitz survives the Holocaust but discovers in the course of the book that his parents did not.

In an essay that presses the case for *Austerlitz* as an exemplary instance of the afterlife of Romanticism, James Chandler points out that the novel's formal experimentation stands as Sebald's solution to the difficulty of the historian's task. "[I]f history is necessary to the purposes of human memory," Chandler proposes, "and history needs a form, and if its forms have degenerated into cliché, then the forms of historiography as we know it must be revitalized by rhetorical genre crossing."[9] One might argue further that because the inadequacy of historiography is particularly evident in relation to the Holocaust, often thought to be a limit case in intelligibility, any effort at representing it demands an especially intense and wide-ranging engagement with generic and formal possibilities. Thus *Austerlitz*'s use of Conradian frame narration; its sometimes bewildering juxtaposition of archival documents and photographs with fictional and semi- or quasi-autobiographical prose; its verbal panoramas; its historical set-pieces, such as one character's re-narration of the events of the battle of Austerlitz; its meta-fictional foregrounding of the unsatisfactory nature of that re-narration; and so forth – the list might be expanded considerably and still remain partial.

Among the most surprising genres to make an appearance is natural history. Informally adopted by the family of a fellow student, Gerald Fitzpatrick, the young Austerlitz spends school vacations with them in Andromeda Lodge, a house filled with specimens: "collections of shells, minerals, beetles, and butterflies; slowworms, adders, and lizards preserved in formaldehyde; snail shells and sea urchins, crabs and shrimps, and large herbaria containing leaves, flowers, and grasses" (*A* 83).[10] These collections date from 1869, we learn, when one of the ancestors of the family met Darwin, "then working on his study of the Descent of Man in a rented house" nearby, a meeting that gave rise to a "schism in the Fitzpatrick clan … whereby one of the two sons in every generation abandoned the Catholic faith to become a natural scientist" (*A* 83–84). The lapsed son of the current generation, Gerald's Great-Uncle Alphonso, initiates Austerlitz into the world of biological science:

Although I did not study natural history later, said Austerlitz, many of Great-Uncle Alphonso's botanical and zoological disquisitions have remained in my mind. Only a few days ago I was rereading that passage in Darwin he once showed me, describing a flock of butterflies flying uninterruptedly for several hours ten miles out from the South American coast, when even with a telescope it was impossible to find a patch of empty sky visible between their whirling wings. But I always found what Alphonso told us at that time about the life and death of moths especially memorable, and of all creatures I still feel the greatest awe for them. (*A* 93)[11]

Strange to say, this passage, too, constitutes part of the novel's attempt to represent the unrepresentable. Like other of Austerlitz's intellectual passions – his interest in monumental architecture, for instance, or railway stations – natural history, and Darwinism in particular, bears directly on the catastrophe he first strives to forget and then to remember. For the Nazi obsession with "race," "breeding," and "purity" derived some of its rationale from a virulent brand of Social Darwinist thinking.[12] But there is another, equally oblique but equally illuminating connection, and it has to do with Sebald's redeployment of evolutionary theory's instantiation of the ubiquity of loss. The passage above continues:

In the warmer months of the year one or other of those nocturnal insects [moths] quite often strays indoors from the small garden behind my house. When I get up early in the morning, I find them clinging to the wall, motionless. I believe, said Austerlitz, they know they have lost their way, since if you do not put them out again carefully they will stay where they are, never moving, until the last breath is out of their bodies, and indeed they will remain in the place where they came to grief even after death, held fast by the tiny claws that stiffened in their last agony, until a draft detaches them and blows them into a dusty corner. Sometimes, seeing one of these moths that have met their end in my house, I wonder what kind of fear and pain they feel while they are lost. (*A* 93–94)

Via Darwin and Alphonso, Austerlitz comes to see other creatures as related to him, fellow travelers in a world of suffering. The fear and pain Austerlitz attributes to the lost moths reflect his own. Whence, I think, the frame narrator's comment not much further along in the novel: "When Austerlitz had brought the tea tray in and was holding slices of white bread on a toasting fork in front of the blue gas flames, I said something about the incomprehensibility of mirror images" (*A* 119). Not only is Austerlitz's loss incomprehensible, so is that of the moths – and so, especially, is the fact that they are reflections of one another. For what exactly does one see in this sort of mirror?

To turn from Sebald back to Darwin is to discern a version of the same question and the same uncertainty about its answer. In "A Biographical Sketch of an Infant" (1877), an account of the early years of his first child drawn from a set of observations written down between 1839 and 1841, Darwin invokes two occasions of self-reflection:

When four and a half months old, he repeatedly smiled at my image and his own in a mirror ... Like all infants he much enjoyed thus looking at himself, and in less than two months perfectly understood that it was an image ... The higher apes which I tried with a small looking-glass behaved differently; they placed their hands behind the glass, and in doing so showed their sense, but far from taking pleasure in looking at themselves they got angry and would look no more.[13]

The mirror anecdotes appear under the subheading *"Association of Ideas, Reason, &c."* Accordingly, that Darwin's son recognizes what appears in the mirror as an image is taken to demonstrate a high degree of understanding. Apes, similarly, show reasoning or "sense" when they reach into the space behind the glass. The article in its entirety constitutes one of the earliest attempts at developmental psychology informed by evolutionary theory, and, as always, Darwin's emphasis falls on human continuity with other animals. But there is difference here as well, and it has to do with divergent affective responses: the human child's pleasure, the apes' anger. Darwin knew of the latter response because he himself had tried the experiment. In the so-called M notebook, kept from July to October, 1838, he records the results of presenting a mirror to an orangutan named Jenny in the London Zoological Gardens: "The young Ourang in <Zoolog>> Gardens *pouts*. partly out displeasure (& partly out of I do not know what when it looked at the glass)."[14]

An uncanny anticipation of the *London Sketch Book* caricature, the scene adumbrated within the parentheses takes for granted what the caricature ridicules: the close connection between apes, and by extension all other animals, and humans. To show Jenny a mirror is already to treat her as a being potentially capable of recognizing herself, reflecting on herself. To admit that one cannot fathom her reaction to such a reflection is to acknowledge her subjectivity, akin to but not the same as one's own. That Darwin stands thus revealed as a key precursor of animal studies should produce no surprise. His difference, however, is instructive, for while the question of the animal as currently posed by thinkers as diverse as Giorgio Agamben, Donna Haraway, and Cary Wolfe repeats both halves of nineteenth-century evolutionary theory's dual query about the human-as-animal and the animal-as-human, it elides what had been central.[15] Such elision attests to our different historical moment but also to a kind of amnesia in connection with that figure (and more than figure) the Victorians suspected stood between themselves and other animals, joining and separating at once. But Darwin, as I have insisted throughout this book, was incapable of forgetting the savage: "compare, the Fuegian & Ourang & outang," he writes in the M notebook, thinking of Jenny again, and of landfall in Tierra del Fuego in 1832, "& dare to say difference so great."[16]

This returns us to "Prof. Darwin." Earlier I considered the possibility that the mirror functions prophetically, giving back an image of the monkey's human future. But it may now be suggested that the mirror reflects not the future but the past, delivers up not a prophecy but a memory. Moths reflect a memory that Austerlitz at first does not realize he possesses and must work to recall; Darwin's son and the orangutan recognize that mirrors give back images, but precisely what they see or why they react as they do cannot be established. The mirror in the *London Sketch Book* caricature, I want to propose, similarly reflects something heretofore invisible and still not directly shown: neither Darwin's face nor the monkey's, but what Victorian naturalists thought of as in between the two, shared by the two. To entertain that proposition is to imagine that the mirror reflects neither of the figures in the caricature but rather, angled to the space between, is empty, or, like the frontispiece to Kingsley's *At Last*

(1871), filled only with palm trees. That emptiness, or that possibly South American but certainly tropical scene, serves as a reminder of what was thought to hold the place between humans and apes, and thus between humans and all other animals: the "missing link," which, writes Gillian Beer, "is an absence, a wide door, a gulf, as well as a cross-over, a chiasmus or creature."[17] As at the end of the *Descent of Man*, here the absence that stands between, that connects but also keeps separate, that remains in memory – that absence bears the name "the savage."

The argument of *Darwin and the Memory of the Human* itself occupies the space between two scenes of reflection: Darwin's "upon first seeing a party of Fuegians on a wild and broken shore" that "such were our ancestors" and Hudson's that "by looking into the empty cavity of one of those broken unburied skulls [he would be] able to see, as in a magic glass, … an image of the world as it once existed in the living brain."[18] Adam Phillips writes about Darwin that he presents an image of humans "as though people were the animals that were haunted by their own and other people's absences."[19] But this seems truer of Sebald. Of Darwin, and of Wallace, Kingsley, and Hudson, we might say instead that they depict people as though they were the animals haunted by the loss of what they imagine comes between them and other animals – "savages" – and as though the only way to free themselves from such haunting were to remember that that loss makes them who they are.

Notes

1. Grant Allen, "Introduction," *In the Guiana Forest: Studies of Nature in Relation to the Struggle for Life*, by James Rodway (London: T. Fisher Unwin, 1894), xiv.
2. *Ibid.*, xv.
3. Clearly, the term "savage" denotes and connotes many things with which one wants to disagree. For that reason, I often place the term, as here, in quotation marks. But such a proceeding allows for its own kind of complacency – as if typography freed one from implication in this or any other living tradition of taxonomic violence. Further, although I find myself using the various alternatives to "savage" from time to time, particularly when I am concerned to distance myself from the objects of my study, those terms are themselves problematic. First peoples, indigenes, indigenous peoples, native peoples, Amerindians: most suggest some geographic primacy in relation to European colonists. But the story evolutionary theory tells of the history of human and proto-human life on the planet reveals a series of migrations, conquests, intermixings, and returns that makes "firstness" dubious both in itself and as a political criterion for an authentic or proper relation to a particular place. Finally, and for me perhaps most importantly, my study takes as one of its central concerns the conceptual, political, and autobiographical work the word "savage" allowed the figures under scrutiny to perform – as well as the sometimes radical transformation of the signification of that word their theories demanded, a transformation that, in the case of Alfred Russel Wallace and W. H. Hudson, amounts to a reversal of the valorization of the poles "civilization" and "savagery." As I hope to show, such reversal was by no means without its own problematic complexities. But putting quotation marks around, placing under erasure, or simply eliding its key term does little to reduce or mitigate those or related difficulties.
4. In connection with my contention as to the status of Victorian South America as a *lieu de mémoire*, see Neil L. Whitehead's description of European travel writing on the continent as "filled with the discovery of the fantastic, the survival of the anachronistic, and the promise of marvelous monstrosity." "South America/Amazonia: The Forest of Marvels," in *The Cambridge Companion to*

Travel Writing, ed. Peter Hulme and Tim Youngs (Cambridge: Cambridge University Press, 2003), 123.

5. Pierre Nora, "Between Memory and History: *Les Lieux de Mémoire*," *Representations* 26 (1989): 7, 12.

6. *Ibid.*, 19.

7. Thomas Henry Huxley, *T. H. Huxley's Diary of the Voyage of H. M. S. Rattlesnake*, ed. Julian Huxley (Garden City, NY: Doubleday, 1931), 20. The statement quoted is from the diary entry for February 7, 1847. A much more elaborate and detailed speculation as to the viability of nineteenth-century colonization in South America appears in a letter written by Richard Spruce in December 1857; see R. E. Schultes, "Richard Spruce and the Potential for European Settlement of the Amazon: An Unpublished Letter," *Botanical Journal of the Linnean Society* 77 (September 1978): 131–39. A botanist with a special interest in the indigenous use of psychoactive plants, Spruce spent a decade and a half in South America (1849–64). The account of his travels, *Notes of a Botanist on the Amazon and Andes*, was published posthumously, and then only because of the editorial efforts of Alfred Russel Wallace, who writes: "I have myself so high an opinion of my friend's work, both literary and scientific, that I venture to think the present volumes will take their place among the most interesting and instructive books of travel of the nineteenth century." Alfred Russel Wallace, "Preface," in *Notes of a Botanist on the Amazon and Andes*, by Richard Spruce, 2 vols. (London: Macmillan, 1908), II: vii.

8. On the advent of stratigraphy as historiography, see Stephen Jay Gould, *Time's Arrow, Time's Cycle: Myth and Metaphor in the Discovery of Geological Time* (Cambridge, MA and London: Harvard University Press, 1987), Noah Heringman, *Romantic Rocks, Aesthetic Geology* (Ithaca, NY and London: Cornell University Press, 2004), and Virginia Zimmerman, *Excavating Victorians* (Albany: State University of New York Press, 2008).

9. On the comparative method, see George Stocking, *Victorian Anthropology* (New York and London: Free Press, 1987) and *Race, Culture, and Evolution: Essays in the History of Anthropology* (Chicago: University of Chicago Press, 1968).

10. Charles Darwin, *The Descent of Man, and Selection in Relation to Sex*, 2 vols. (1871; reprint, Princeton, NJ: Princeton University Press, 1981), II: 405. Given the weight the rest of the *Descent* places on the close connection between human mental abilities and predilections and those of other animals, something to which I will give considerable attention in the first chapter of this book, it is remarkable that this final sentence mentions physical attributes only. But Gowan Dawson suggests the degree to which this may have been a strategic misrepresentation on Darwin's part given the particularly fierce opposition among his critics to the "assertion that the human capacity for moral action" was, like the human form, the product of natural and sexual selection. *Darwin, Literature and Victorian Respectability* (Cambridge and New York: Cambridge University Press, 2007), 28.

11. John Frow, *Time and Commodity Culture: Essays in Cultural Theory and Postmodernity* (Oxford: Clarendon, 1997), 230.

12. See Bruno Latour, *We Have Never Been Modern* (Cambridge, MA: Harvard University Press, 1993). Dana Seitler treats another significant episode in the production of the human by way of an exchange between purity and hybridity in *Atavistic Tendencies: The Culture of Science in American Modernity* (Minneapolis and London: University of Minnesota Press, 2008).

13. Many accounts of that formation exist. See David Amigoni, *Colonies, Cults, and Evolution: Literature, Science and Culture in Nineteenth-Century Writing* (Cambridge and New York: Cambridge University Press, 2007); Peter Bowler, *Fossils and Progress: Paleontology and the Idea of Progressive Evolution in the Nineteenth Century* (New York: Science History Publications, 1976), *The Invention of Progress: The Victorians and the Past* (Oxford and New York: Blackwell, 1990), and *Theories of Human Evolution: A Century of Debate, 1844–1944* (Baltimore, MD: Johns Hopkins University Press, 1986); Donald Grayson, ed., *The Establishment of Human Antiquity* (New York: Academic, 1983); Barry M. Marsden, *Pioneers of Prehistory: Leaders and Landmarks in English Archaeology, 1500–1900* (Ormskirk: Hesketh, 1984); and A. Bowdoin Van Riper, *Men Among the Mammoths: Victorian Science and the Discovery of Human Prehistory* (Chicago: University of Chicago Press, 1993). On evolutionary thought and race, see Stephen Jay Gould, *The Mismeasure of Man* (New York: Norton, 1981); Stocking, *Victorian Anthropology* and *Race, Culture, and Evolution*; and Robert J. C. Young, *Colonial Desire: Hybridity in Theory, Culture, and Race* (London and New York: Routledge, 1995).

14. Discussing the notion of the human-as-animal as self-evident or commonplace, Barbara Herrnstein Smith notes: "*Of course* we are animals, it is said … – the straightforward truth here deriving, presumably, from evolutionary theory and the current scheme of biological classification. It's not always clear, however, that the classifications and distinctions of empirical science should be awarded such unproblematic ontological, not to mention epistemic or, especially, ethical authority." "Animal Relatives, Difficult Relations," *differences* 15.1 (2004): 2.

15. Michel Foucault illustrates an earlier shift in the practice of natural history, a shift he characterizes, with one of his inimitable paradigmatic anecdotes, as the advent of historicity in the study of nature:

> One day, towards the end of the eighteenth century, [Georges] Cuvier was to topple the glass jars of the Museum, smash them open and dissect all the forms of animal visibility that the Classical age had preserved in them … [This act] was also to be the beginning of what, by substituting anatomy for classification, organism for structure, internal subordination for visible character, the series for tabulation, was to make possible the precipitation into the old flat world of animals and plants, engraved in black and white, a whole profound mass of time to which men were to give the renewed name of *history*. (*The Order of Things: An Archaeology of the Human*

Sciences (1966; English translation, New York: Pantheon, 1970), 137–38; emphasis in the original here and throughout unless otherwise noted)

16. Such catholicity sometimes implied, but sometimes did not, that humans ought to be considered part of the natural world. The representative pre-Victorian positions are treated by Emma Spary, "Political, Natural and Bodily Economies," in *Cultures of Natural History*, ed. N. Jardine, J.A. Secord, and Spary (Cambridge: Cambridge University Press, 1996), 178–96, and Paul B. Wood, "The Science of Man," in *Cultures of Natural History*, 197–210.

17. On the traffic in human body parts integral to the workings of European racial science, see Tim Fulford, Debbie Lee, and Peter J. Kitson, *Literature, Science and Exploration in the Romantic Era: Bodies of Knowledge* (Cambridge and New York: Cambridge University Press, 2004), especially chapter 6, "Exploration, Headhunting and Race Theory: The Skull Beneath the Skin."

18. Nora, "Between Memory and History," 15.

19. Canning quoted in William W. Kaufmann, *British Policy and the Independence of Latin America, 1804–1828* (New Haven, CT: Yale University Press, 1951), 178.

20. Leslie Bethell, "Britain and Latin America in Historical Perspective," in *Britain and Latin America: A Changing Relationship*, ed. Victor Bulmer-Thomas (Cambridge and New York: Cambridge University Press, 1989), 1.

21. Relations between Victorian Britain and Latin America have received relatively little scholarly attention to date, especially in literary, cultural, and science studies. A small but growing number of exceptions include Robert Aguirre, *Informal Empire: Mexico and Central America in Victorian Culture* (Minneapolis and London: University of Minnesota Press, 2005); Robert Aguirre and Ross Forman, eds., *Connecting Continents: Britain and Latin America, 1780–1900* (New York and Amsterdam: Rodopi, forthcoming); D. Graham Burnett, *Masters of All They Surveyed: Exploration, Geography, and a British El Dorado* (Chicago and London: University of Chicago Press, 2000); Mary Louise Pratt, *Imperial Eyes: Travel Writing and Transculturation* (London and New York: Routledge, 1992); and Nancy Leys Stepan, *Picturing Tropical Nature* (Ithaca, NY: Cornell University Press, 2001).

22. Anne McClintock, *Imperial Leather: Race, Gender, and Sexuality in the Colonial Contest* (New York and London: Routledge, 1995), 40.

23. Arthur Conan Doyle, *The Lost World* (1912), reprinted in *The Lost World and Other Stories* (Ware: Wordsworth, 1995), 50.

24. *Ibid.*, 27.

25. Everard F. im Thurn, *Among the Indians of Guiana, being Sketches Chiefly Anthropologic from the Interior of British Guiana* (London: Kegan, Paul, Trench and Co., 1883), 85. Rosamund Dalziel elucidates the relation between im Thurn's work and Doyle's novel in "The Curious Case of Sir Everard im Thurn and Sir Arthur Conan Doyle: Exploration and the Imperial Adventure Novel, *The Lost World*," *ELT* 45.2 (2002): 131–57.

26. See John Gallagher and Ronald Robinson, "The Imperialism of Free Trade," *Economic History Review* 6 (1953): 1–15; D. C. M. Platt, ed., *Business Imperialism, 1840–1930: An Inquiry Based on British Experience in Latin America* (Oxford: Clarendon, 1977).

27. P. J. Cain and A. G. Hopkins, *British Imperialism: Innovation and Expansion 1688–1914* (London and New York: Longman, 1993), 276.

28. *Ibid.*, 42–43. Cain and Hopkins address specifically the causes of British imperial expansion, arguing: "It was spearheaded, not by manufacturing interests, but by gentlemanly elites who saw in empire a means of generating income flows in ways that were compatible with the high ideals of honour and duty" (46). For the continuing debate on this thesis, see H. V. Bowen, *The Business of Empire: The East India Company and Imperial Britain, 1756–1833* (Cambridge and New York: Cambridge University Press, 2006); the contributions to Raymond E. Dumett, ed., *Gentlemanly Capitalism and British Imperialism: The New Debate on Empire* (New York: Longman, 1999); and Robert Freeman Smith, "Latin America, the United States and the European Powers, 1830–1930," in *The Cambridge History of Latin America*, ed. Leslie Bethell, 11 vols. (Cambridge: Cambridge University Press, 1984–94), iv: 83–119.

29. Stepan, *Picturing Tropical Nature*, 19.

30. Roderick Nash makes the case that nature-appreciation itself can be considered a commodity; see "The Export and Import of Nature," *Perspectives in American History* 12 (1979): 519–60. Kavita Philip details the politically and practically delicate undertaking of relocating cinchona trees, the source of quinine, from the slopes of the Andes to Kew Gardens and then on to the hill plantations of British-held India in "Imperial Science Rescues a Tree: Global Botanic Networks, Local Knowledge and the Transcontinental Transplantation of Cinchona," *Environment and History* 1 (1995): 173–200.

31. Pratt, *Imperial Eyes*, 7, 33. For two particularly valuable treatments of these issues, see Richard Drayton, *Nature's Government: Science, Imperial Britain, and the "Improvement" of the World* (New Haven, CT and London: Yale University Press, 2000), and Jim Endersby, *Imperial Nature: Joseph Hooker and the Practices of Victorian Science* (Chicago and London: University of Chicago Press, 2008). On related questions, see also Fa-ti Fan, *British Naturalists in Qing China: Science, Empire, and Cultural Encounter* (Cambridge, MA and London: Harvard University Press, 2004), and Beth Fowkes Tobin, *Colonizing Nature: The Tropics in British Arts and Letters, 1760–1820* (Philadelphia: University of Pennsylvania Press, 2005).

32. D. Graham Burnett, "'It Is Impossible to Make a Step without the Indians': Nineteenth-Century Geographical Exploration and the Amerindians of British Guiana," *Ethnohistory* 49.1 (Winter 2002): 6. Another useful treatment of the complicity of Victorian science with empire is Harriet Ritvo, "Zoological Nomenclature and the Empire of Victorian Science," in *Victorian Science in Context*, ed. Bernard Lightman (Chicago: University of Chicago Press, 1997), 334–53.

33. Aguirre, *Informal Empire*, xvi.

34. Edward Said, *Culture and Imperialism* (New York: Knopf, 1993), 52.

35. Jonathan Lamb, *Preserving the Self in the South Seas, 1680–1840* (Chicago and London: University of Chicago Press, 2001), 7.

36. Jonathan Lamb, "Metamorphosis and Settlement: The Enlightened Anthropology of Colonial Societies," in *The Anthropology of the Enlightenment*, ed. Larry Wolff and Marco Cipolloni (Stanford, CA: Stanford University Press, 2007), 289.

37. This is, in brief, the argument of Hudson's first novel, *The Purple Land that England Lost* (1885), a *roman à thèse* masquerading as a collection of picaresque tales.

38. Charles Darwin, *Journal of Researches into the Geology and Natural History of the Various Countries Visited by H.M.S. Beagle, under the Command of Captain FitzRoy, R.N, from 1832 to 1836* (1839), vols. II and III, *The Works of Charles Darwin*, 29 vols. (London: Pickering, 1986–89), II: 10–11.

39. For a more extended consideration of affective epistemology in the work of Darwin and Wallace, see my "Victorian Beetlemania," in *Victorian Animal Dreams: Representations of Animals in Victorian Literature and Culture*, ed. Deborah Morse and Martin Danahay (Aldershot, Hampshire, and Burlington, VT: Ashgate, 2007), 35–51.

40. Charles Kingsley, *At Last: A Christmas in the West Indies*, volume XIV of *The Works of Charles Kingsley*, 28 vols. (London: Macmillan, 1880–85), 1.

41. Spaniards and Portuguese themselves, however, conducted numerous important natural-historical investigations. For an account of expeditions to Mexico, Central America, and South America mounted by the Spanish Bourbons, see Angel Guirao de Vierna, "Análisis cuantitativo de las expediciones españolas con destino al Nuevo Mundo," in *Ciencia, vida y espacio en Iberoamérica*, ed. José Luis Pestet 3 vols. (Madrid: Consejo Superior de Investigaciones Científicas, 1989), III: 65–93.

42. Nineteenth-century English translations of the works of these travelers include Spix and Martius, *Travels in Brazil, in the years 1817–20* (1824) and Adalbert, *Travels of His Royal Highness Prince Adalbert of Prussia in the south of Europe and in Brazil, with a voyage up the Amazon and Xingu* (1849). For full publication details on these and others, see "Outstanding Nineteenth-Century Translations," the appendix to Bernard Naylor, *Accounts of Nineteenth-Century South America* (London: Athlone, 1969).

43. On the role of Humboldt, see, among others, William H. Brock, "Humboldt and the British: A Note on the Character of British Science," *Annals of Science 50* (Basingstoke: Taylor and Francis, 1993): 365–72; Susan Faye Canon, *Science in Culture: The Early Victorian Period* (New York: Science History Publications, 1978); Nigel Leask, "Darwin's 'Second Sun': Alexander von Humboldt and the Genesis of *The Voyage of the Beagle*," in *Literature, Science Psychoanalysis, 1830–1970: Essays in Honor of Gillian Beer*, ed. Helen Small and Trudi Tate (Oxford and New York: Oxford University Press, 2003), 13–36; Nicholas Rupke, "A Geography of Enlightenment: The Critical

Reception of Humboldt's Work," in *Geography and Enlightenment*, ed. David N. Livingstone and Charles Withers (Chicago: University of Chicago Press, 1999), 319–39; Pratt, *Imperial Eyes*; Stepan, *Picturing Tropical Nature*; and Victor Wolfgang Von Hagen, *South America Called Them: Explorations of the Great Naturalists La Condamine, Humboldt, Darwin, Spruce* (New York: Knopf, 1945).

44. Darwin, *Journal*, 473.
45. In *The Structure of Scientific Revolutions* (1962; third edition, Chicago and London: University of Chicago Press, 1996), Thomas Kuhn contended that the history of science could be described as relatively quiescent periods of "normal science" occasionally punctuated by paradigm-shifting scientific revolutions.
46. In *Belated Travelers*, Ali Behdad anatomizes "the sense of belatedness" experienced by European travelers to the Orient "at a time when the European colonial power structure and the rise of tourism had transformed the exotic referent into the familiar sign of Western hegemony." For the writers Behdad considers, the effects of that sense are differential; nevertheless, belatedness always falls under the sign of a loss, and a loss in relation specifically to a "*desire for the Orient.*" *Belated Travelers: Orientalism in the Age of Colonial Dissolution* (Durham, NC and London: Duke University Press, 1994), 13–14. But for Victorian naturalists traveling in and writing about South America, belatedness figures as something both lost and gained in relation to signal precursors and "savage" ancestors.
47. Burnett, *Masters of All They Surveyed*, 39, 47.
48. The phrase "man-like apes" is T. H. Huxley's, from *Evidence as to Man's Place in Nature* (London and Edinburgh: Williams and Norgate, 1863), 1.
49. Jardine and Spary, "Natures of Cultural History," in *Cultures of Natural History*, 12.
50. Janet Browne, *Charles Darwin: Voyaging* (New York: Knopf, 1995), 207.
51. A. H. G. Alston, "Henry Walter Bates: A Centenary," *The Geographical Journal* 112 (July–September 1948), 2. Alex Shoumatoff, in his introduction to the Penguin edition of Bates's *The Naturalist on the River Amazons*, glosses Batesian mimicry as "a palatable species mimicking an unpalatable one." "Introduction," *The Naturalist on the River Amazons*, by Henry Walter Bates (1863; reprint, Harmondsworth: Penguin, 1989), xiii.
52. Henry Walter Bates, Pocket-Book kept while in Amazon (1848–59), BL Add. MS 42138A-B, folios 24–25.
53. Shoumatoff, "Introduction," xii.
54. On Stevens and the role of personal relationships in natural historical collecting during the mid-Victorian period, see Jane Camerini, "Wallace in the Field," *Osiris*, second series, 11 (1996): 44–65, and "Remains of the Day: Early Victorians in the Field," in *Victorian Science in Context*, ed. Bernard Lightman (Chicago: University of Chicago Press, 1997) 354–77. Judith Pascoe's *The Hummingbird Cabinet: A Rare and Curious History of Romantic Collectors* (Ithaca, NY and London: Cornell University Press, 2005) provides

a fascinating account of natural-historical and other collecting in the Romantic period.

55. On nineteenth-century natural history, see David Elliston Allen, *The Naturalist in Britain: A Social History* (1976; reprint, Harmondsworth: Penguin, 1978); Lynn Barber, *The Heyday of Natural History, 1820–1870* (London: Cape, 1980); Lynn L. Merrill, *The Romance of Victorian Natural History* (New York: Oxford University Press, 1989); Peter Raby, *Bright Paradise: Victorian Scientific Travellers* (London: Chatto and Windus, 1996); Barbara T. Gates, *Kindred Nature: Victorian and Edwardian Women Embrace the Living World* (Chicago: University of Chicago Press, 1999); and Jardine, Secord, and Spary, eds., *Cultures of Natural History*. The potentially disastrous consequences of the imperative to collect were already beginning to be understood. As Edmund Gosse writes in *Father and Son*:

> The ring of living beauty drawn around our [British] shores was a very thin and fragile one. It had existed all those centuries solely in consequence of the indifference, the blissful ignorance of man. These rock-basins, fringed by corallines, filled with still water almost as pellucid as the upper air itself, thronged with beautiful sensitive forms of life, – they exist no longer, they are all profaned, and emptied, and vulgarized. An army of 'collectors' has passed over them, and ravaged every corner of them. (*Father and Son* (1907; reprint, Harmondsworth: Penguin, 1983), 125)

56. Lorraine Daston treats memory and natural history, and specifically the function of holotypes and stabilized botanical nomenclature in adjudicating between a certain kind of memory and a certain kind of forgetting, in "Type Specimens and Scientific Memory," *Critical Inquiry* 31 (Autumn 2004): 153–82.

57. Alfred Russel Wallace, *A Narrative of Travels on the Amazon and Rio Negro, with an Account of the Native Tribes, and Observations on the Climate, Geology, and Natural History of the Amazon Valley* (London: Reeve and Co., 1853), 154. Robert Schomburgk, although not the first European to see *Victoria regia*, suggested its nationalistic scientific appellation. Peter Rivière, ed., *The Guiana Travels of Robert Schomburgk 1835–1844, Volume 1: Explorations on Behalf of the Royal Geographical Society 1835–1839* (Aldershot, Hampshire, and Burlington, VT: Ashgate, for the Hakluyt Society, 2006), 231.

58. For a fascinating treatment of mass media as "prosthetic memory," see Alison Landsberg, *Prosthetic Memory: The Transformation of American Remembrance in the Age of Mass Culture* (New York: Columbia University Press, 2004).

59. Ann C. Colley, *Nostalgia and Recollection in Victorian Culture* (Basingstoke and London: Macmillan; New York: St. Martin's, 1998), 22.

60. Of the difficulty of narrating discovery, Pratt writes: "The verbal painter must render momentously significant what is, especially from a narrative point of view, practically a non-event." *Imperial Eyes*, 202. For a now classic redescription of the process of scientific discovery, see the sixth chapter of Kuhn, *The Structure of Scientific Revolutions*.

61. Henry Walter Bates, *The Naturalist on the River Amazons, A Record of Adventures, Habits of Animals, Sketches of Brazilian and Indian Life, and*

Aspects of Nature under the Equator, During Eleven Years of Travel, 2 vols. (London: John Murray, 1863), II: 415.

62. Allen, "Introduction," xi; my emphasis.

63. *Ibid.*, vii–viii. Nancy Rose Marshall addresses the characteristic conflation of tropicality, archaism, and struggle in "'A Dim World, Where Monsters Dwell': The Spatial Time of the Sydenham Crystal Palace Dinosaur Park," *Victorian Studies* 49.2 (Winter 2007): 286–301.

64. Charles Darwin, *On the Origin of Species by Means of Natural Selection, or the Preservation of Favoured Races in the Struggle for Life* (1859; reprint, Cambridge, MA and London: Harvard University Press, 1964), 489–90.

65. Allen, "Introduction," xiv; my emphasis.

66. *Ibid.*, xv, xvii.

67. Compare Sarmiento's earlier diagnosis of what he perceives to be the cause of Argentina's violent civil conflicts:

> Two distinct forms of civilization meet upon a common ground in the Argentine Republic: one, still in its infancy, which, ignorant of that so far above it, goes on repeating the crude efforts of the Middle Ages; the other, disregarding what lies at its feet, while it strives to realize itself in the latest results of European civilization; the nineteenth and twelfth centuries dwell together – one inside the cities, the other without them. (Domingo Faustino Sarmiento, *Life in the Argentine Republic in the Days of the Tyrants; or, Civilization and Barbarism*, trans. Mary Mann (first edition, 1845; English translation, from the third Spanish edition, New York: Hurd and Houghton; Cambridge: Riverside Press, 1868), 42)

68. Allen, "Introduction", x. Compare Rodway, who devotes the second chapter of *In the Guiana Forest* to his memories and estimation of "The Man of the Forest." At first the appropriateness of the term "savage" itself is denied: "If to be a savage means to be rude and uncouth, ill-mannered and disagreeable, then the Indian little deserves such an appellation. He is one of nature's gentlemen." But the emphasis in the phrase "nature's gentlemen" seems to fall on "nature," for it quickly becomes apparent that Rodway understands the indigenous inhabitants of Guiana to exist on the border between human and non-human: "Nowhere perhaps is the fauna of such an ancient type, so well protected, and so perfectly fitted to its environment, and nowhere can we study man as an animal so well as in the Guiana forest." Rodway, *In the Guiana Forest*, 17, 19.

69. Rodway, *In the Guiana Forest*, 11. Spencer first uses the phrase "survival of the fittest" in *The Principles of Biology*: "This survival of the fittest, which I have here sought to express in mechanical terms, is that which Mr. Darwin calls 'natural selection.'" Herber Spencer, *The Principles of Biology*, 2 vols. (London: Williams and Norgate, 1864), I: 444–45.

70. Rodway, *In the Guiana Forest*, 13.

71. *Ibid.*, 234–35.

72. *Ibid.*, 239. Although I note Rodway's closer proximity to Lamarck than Darwin here, it is a mistake to view the two as starkly opposed on this matter. Darwin never entirely abandoned the notion that inheritance of

acquired characteristics played some role in evolutionary change. In the late "A Biograpical Sketch of an Infant," for instance, he writes: "May we not suspect that the vague but very real fears of children, which are quite independent of experience, are the inherited effects of real dangers and abject superstitions during ancient savage times?" Charles Darwin, "A Biographical Sketch of an Infant," *Mind* 2 (July 1877): 288. He did, however, insist that the role of inherited acquired characteristics was a small one when compared to that of natural and sexual selection. For an interesting treatment of the powerful afterlife of Lamarckian ideas about inheritance, see Shafquat Towheed, "The Creative Evolution of Scientific Paradigms: Vernon Lee and the Debate over the Hereditary Transmission of Acquired Characteristics," *Victorian Studies* 49.1 (Autumn 2006): 33–61.

73. Rodway, *In the Guiana Forest*, 239, 241, 13, 240.

74. Walter Bagehot, *Physics and Politics; or, Thoughts on the Application of the Principles of "Natural Selection" and "Inheritance" to Political Society* (New York: D. Appleton and Company, 1873), 2–3.

75. Laura Otis, *Organic Memory: History and the Body in the Late Nineteenth and Early Twentieth Centuries* (Lincoln: University of Nebraska Press, 1994), 13–23. In *Unconscious Memory*, Samuel Butler attempts to restore teleology to evolutionary change by granting memory the role both of preserving organisms intact from one generation to the next and providing the impetus for change. About the relation between individuals of different generations he asserts "the oneness of personality between parents and offspring – memory on the part of offspring of certain actions which it did when in the persons of its forefathers; the latency of that memory until it is rekindled by a recurrence of associated ideas, and the unconsciousness with which habitual actions come to be performed." *Unconscious Memory* (1880; reprint, New York: Dutton, 1911), 19.

76. Otis, *Organic Memory*, 3. On bodily memory in the nineteenth century, see also Sally Shuttleworth, "'The Malady of Thought': Embodied Memory in Victorian Psychology and the Novel," in *Memory and Memorials, 1789–1914: Literary and Cultural Perspectives*, ed. Matthew Campbell, Jacqueline M. Labbé, and Shuttleworth (London and New York: Routledge, 2000), 46–59. Grant Allen based some of his fiction on the possibility of such memory; see, for instance, "A Freak of Memory," *The Queen, the Lady's Newspaper* 102 (November 13, 1897): 909–11.

77. Gillian Beer, *Open Fields: Science in Cultural Encounter* (Oxford: Clarendon, 1996), 143.

78. But see note 72, above.

79. Darwin, *Descent*, 1: 32.

80. In M notebook, kept between July and October 1838, Darwin writes: "Now if memory <<of a tune & words>> can thus lie dormant, during a whole life time, quite unconsciously of it, surely memory from one generation to another, also without consciousness, as instincts are, is not so very wonderful." *Charles Darwin's Notebooks, 1836–1844: Geology, Transmutation*

of Species, Metaphysical Enquiries, ed. Paul H. Barrett *et al.* (Ithaca, NY: Cornell University Press, 1987), 521.

81. In this regard, Nora owes something to Maurice Halbwachs; see his *On Collective Memory*, trans. Lewis Coser (Chicago: University of Chicago Press, 1992).

82. Terdiman's book contributes to an ongoing discussion of memory, the bibliography for which is too enormous to be recapitulated here in any but the most cursory terms. Foundational accounts include Henri Bergson, *Matter and Memory*, trans. Nancy Margaret Paul and W. Scott Palmer (1896; fifth edition, 1908; English translation, New York: Zone Books, 1991); Halbwachs, *On Collective Memory*; William James, *The Principles of Psychology*, 2 vols. (1890; reprint, Cambridge, MA: Harvard University Press, 1981); Frances Yates, *The Art of Memory* (Chicago: University of Chicago Press, 1966). For more recent contributions, see Ali Behdad, *A Forgetful Nation: On Immigration and Cultural Identity in the United States* (Durham, NC and London: Duke University Press, 2005); Mary Carruthers, *The Book of Memory: A Study of Memory in Medieval Culture* (Cambridge: Cambridge University Press, 1990); Paul Connerton, *How Societies Remember* (Cambridge and New York: Cambridge University Press, 1989); Andreas Huyssen, *Present Pasts: Urban Palimpsests and the Politics of Memory* (Stanford, CA: Stanford University Press, 2003) and *Twilight Memories: Marking Time in a Culture of Amnesia* (London: Routledge, 1995); Jacques Le Goff, *History and Memory*, trans. Steven Rendall and Elizabeth Claman (New York: Columbia University Press, 1992); Ann Rosalind Jones and Peter Stallybrass, *Renaissance Clothing and the Materials of Memory* (Cambridge and New York: Cambridge University Press, 2000); Nora, "Between Memory and History"; Paul Ricoeur, *Memory, History, Forgetting*, trans. Kathleen Blamey and David Pellauer (Chicago and London: University of Chicago Press, 2004); Elizabeth Kowaleski Wallace, *The British Slave Trade and Public Memory* (New York: Columbia University Press, 2006). Trauma studies has had a particular interest in retheorizing memory; see especially the essays in Cathy Caruth, ed., *Trauma: Explorations in Memory* (Baltimore, MD: Johns Hopkins University Press, 1995). In "Trauma, Memory, and the Railway Disaster: The Dickensian Connection," *Victorian Studies* 43.3 (Spring 2001): 413–36, Jill Matus discusses the advent of the concept of psychic trauma in the Victorian period. For other notable investigations of nineteenth-century memory, see Gillian Beer, "Origins and Oblivion in Victorian Narrative," in *Sex, Politics, and Science in the Nineteenth-Century Novel*, ed. Ruth Bernard Yeazell (Baltimore, MD: Johns Hopkins University Press, 1986), 63–87; Chris Bongie, *Exotic Memories: Literature, Colonialism, and the Fin de Siècle* (Stanford, CA: Stanford University Press, 1991); Colley, *Nostalgia and Recollection*; Nicholas Dames, *Amnesiac Selves: Nostalgia, Forgetting, and British Fiction, 1810–1870* (Oxford and New York: Oxford University Press, 2001); the essays in Matthew Campbell, Jacqueline M. Labbé, and Sally Shuttleworth, eds., *Memory and Memorials, 1789–1914: Literary and Cultural*

Perspectives (London and New York: Routledge, 2000); Frances Ferguson, "Romantic Memory," *Studies in Romanticism* 35 (1996): 509–33; and Otis, *Organic Memory*.

83. Terdiman carefully specifies his interest in "perceptions of a fundamental difference between the nineteenth century's problematic experience of memory and the past it imagined it had lost in gaining what we know as modernity." Richard Terdiman, *Present Past: Modernity and the Memory Crisis* (Ithaca, NY: Cornell University Press, 1993), 38. Other thinkers, including Reinhart Koselleck (see note 86, below), seem to believe such loss actually to have occurred, thus opening themselves up to the objection that Frow levels at Nora's work on *lieux de mémoire*: namely, that it is "nostalgic" in the sense that it is structured "by a series of contradictions between a realm of authenticity and fullness of being, and the actually existing 'forms of human association' – a contradiction often projected on to a quasi-historical axis as that between modern and traditional societies." *Time and Commodity Culture*, 222; Frow quotes from Bryan S. Turner, "A Note on Nostalgia," *Theory, Culture and Society* 4 (1987): 147–56.

84. Terdiman, *Present Past*, vii.

85. *Ibid.*, 5.

86. Other influential accounts of memory and modernity point to a rupture not so much between past and present as between present and future. Reinhart Koselleck, in *Futures Past: On the Semantics of Historical Time*, argues that constitutive of the modern is a divergence of what he terms the "horizon of expectation" from the "space of experience": "My thesis is that during the *Neuzeit* [modern times] the difference between experience and expectation has increasingly expanded; more precisely, that *Neuzeit* is first understood as a *neue Zeit* [new age] from the time that expectations have distanced themselves evermore from all previous experience." Koselleck advances empirical causes for such distancing, including the technological innovation associated with the Industrial Revolution; the socio-political innovation for which the French Revolution can stand by way of synecdoche; and secularization, which resulted in the disappearance of the "immovable limit to the horizon of expectation" to be found in Christian eschatology. The consequences, while also empirical, are perhaps more interestingly epistemological, for in the absence of a reliable correspondence between experience and expectation the value of the past to the formation of knowledge becomes tenuous. Whereas for Terdiman modernity places increased pressure on memory, for Koselleck memory can appear to be irrelevant, a casualty of the modern. Reinhart Koselleck, *Futures Past: On the Semantics of Historical Time*, trans. Keith Tribe (Cambridge, MA and London: MIT Press, 1985), 276, 277. Jürgen Habermas usefully discusses Koselleck's thesis in relation to Benjamin's and Hegel's conceptions of modernity; see "Modernity's Consciousness of Time," in *The Philosophical Discourse of Modernity: Twelve Lectures*, trans. Frederick Lawrence (Cambridge, MA: MIT Press, 1987), 1–22.

87. Terdiman, *Present Past*, vii.

88. Terdiman's key figures are Alfred Musset, Baudelaire, Proust, and Freud, and only in connection with the last does the question of a science of memory appear. For the outstanding exceptions to my generalization about the attention science has received in studies of changing ideas about and relations to memory, see Ian Hacking, *Rewriting the Soul: Multiple Personality and the Sciences of Memory* (Princeton, NJ: Princeton University Press, 1995) and Otis, *Organic Memory*.

89. Benjaminian "experience" is characteristic of a situation in which "certain contents of the individual past combine with material of the collective past." "On Some Motifs in Baudelaire," in *Illuminations*, trans. Harry Zohn (New York: Schocken, 1968), 159. Jakki Spicer observes that the chronology Benjamin provides here differs from that to be found elsewhere in his work, where the eradication of experience is keyed to a much earlier event, the rise of the novel; "The Author is Dead, Long Live the Author: Autobiography and the Fantasy of the Individual," *Criticism* 47.3 (Summer 2005): 387–403.

90. Benjamin, "On Some Motifs," 182.

91. *Ibid.*, 157. A notable exception is Miriam Hansen, who, in the process of effecting a powerful resolution to the apparent paradox of Benjamin's attitude to the coming of the modern, celebratory and anxious by turns, details the crucial links between Benjaminian theory and early twentieth-century physiology; "Benjamin and Cinema: Not a One-Way Street," in *Benjamin's Ghosts: Interventions in Contemporary Literary and Cultural Theory*, ed. Gerhard Richter (Stanford, CA: Stanford University Press, 2002), 41–73.

92. Walter Benjamin, "The Storyteller," in *Illuminations*, 94.

93. Charles Lyell, *Principles of Geology, Being an Attempt to Explain the Former Changes of the Earth's Surface, by Reference to Causes Now in Operation*, 3 vols. (1830–33; reprint, Chicago: University of Chicago Press, 1991), III: 384. Darwin, discussing the evidence geology provides of a nearly unfathomable lapse of time, writes: "The consideration of these facts impresses my mind almost in the same manner as does the vain endeavour to grapple with the idea of eternity." *Origin of Species*, 285.

94. The debate is recounted in many places. Here I borrow from the late nineteenth-century retrospect cited by Robert J. Richards in *Darwin and the Emergence of Evolutionary Theories of Mind and Behavior* (Chicago and London: University of Chicago Press, 1987), 3–4. Dawson demonstrates in *Darwin, Literature and Victorian Respectability* that the association between evolutionary theory and Wilberforce's suggestion of implicitly sexual shame grew more powerful as the century progressed.

95. Ernst Haeckel, *The Evolution of Man: A Popular Exposition of the Principal Points of Human Ontogeny and Phylogeny*, 2 vols. (1874; English translation of third edition, New York: D. Appleton and Co., 1879), I: 12. For the most complete treatment to date of Haeckel in the context of nineteenth-century evolutionary theory, see Robert J. Richards, *The Tragic Sense of Life: Ernst Haeckel and the Struggle over Evolutionary Thought* (Chicago and London: University of Chicago Press, 2008).

96. Robert J. Richards, *The Meaning of Evolution: The Morphological Construction and Ideological Reconstruction of Darwin's Theory* (Chicago and London: University of Chicago Press, 1992), 152. See also Richards's "The Epistemology of Historical Interpretation: Progressivity and Recapitulation in Darwin's Theory," in *Biology and Epistemology*, ed. Richard Creath and Jane Maienschein (Cambridge and New York: Cambridge University Press, 2000), 64–88, and, for the position against which Richards argues, Stephen Jay Gould, *Ontogeny and Phylogeny* (Cambridge, MA and London: Harvard University Press, 1977).

97. Charles Darwin, *The Autobiography of Charles Darwin, 1809–1882* (1887; reprint, New York and London: Norton, 1969), 21. The first record of his reaction to Fuegians appears in a diary entry written in December of 1832: "It was without exception the most curious spectacle I ever beheld. I would not have believed how entire the difference between savage & civilized man is. It is greater than between a wild & domesticated animal, in as much as in man there is greater power of improvement." Darwin, *Diary of the Voyage of H.M.S. Beagle*, vol. I, *The Works of Charles Darwin*, III.

98. Kingsley, *At Last*, 387.

99. The term is James A. Leith's; see his review of Pierre Nora, ed., *Realms of Memory: the Construction of the French Past*, vol. III, *The Symbols*, English language edition ed. Laurence D. Kritzman, trans. Arthur Goldhammer (New York and Chichester, West Sussex: Columbia University Press, 1998); www.h-france.net/reviews/leith.html.

100. Forbes Phillips, "Ancestral Memory: A Suggestion," *Nineteenth Century* 59 (1906): 981.

101. Phillips may be alluding specifically to the phonograph. See Ivan Kreilkamp's contentions about the implications for memory of the 1877 invention of the phonograph in "A Voice Without a Body: The Phonographic Logic of *Heart of Darkness*," the final chapter of *Voice and the Victorian Storyteller* (Cambridge and New York: Cambridge University Press, 2005), 179–205.

I CHARLES DARWIN'S SAVAGE MNEMONICS

1. William Henry Hudson, *Idle Days in Patagonia*, vol. XVI of *Collected Works of W. H. Hudson*, 24 vols. (London: J. M. Dent; New York: Dutton, 1923), 199; emphasis in the original here and throughout unless otherwise specified.

2. *Ibid.*, 205.

3. *The Naturalist in La Plata*, vol. VIII of *Collected Works of W. H. Hudson*, 379. I treat this episode at more length in the fourth chapter of this book.

4. Johannes Fabian, *Time and the Other: How Anthropology Makes Its Object* (New York: Columbia University Press, 1983), 31; Anne McClintock, *Imperial Leather: Race, Gender, and Sexuality in the Colonial Contest* (New York and London: Routledge, 1995), 30. Fabian defines the

"denial of coevalness" as *"a persistent and systematic tendency to place the referent(s) of anthropology in a Time other than the present of the producer of anthropological discourse"* (*Time and the Other*, 31). As this definition suggests, Fabian concerns himself specifically with anthropologists and the temporal differential they assume exists between themselves and their objects of study. But in this regard anthropology shares an assumption widespread in nineteenth-century European thought about non-European peoples. See also Brad Evans, *Before Cultures: The Ethnographic Imagination in American Literature, 1865–1920* (Chicago and London: University of Chicago Press, 2005), and Brian Street, *The Savage in Literature: Representations of "Primitive" Society in English Fiction 1858–1920* (London and Boston: Routledge and Kegan Paul, 1975), 10.

5. Haeckel: "The evolution of the germ (Ontogeny) is a compressed and shortened reproduction of the evolution of the tribe (Phylogeny)." *The Evolution of Man: A Popular Exposition of the Principal Points of Human Ontogeny and Phylogeny*, 2 vols. (1874; English translation of third edition, New York: D. Appleton and Co., 1879), I: 12. Robert J. Richards, in *The Meaning of Evolution: The Morphological Construction and Ideological Reconstruction of Darwin's Theory* (Chicago and London: University of Chicago Press, 1992), traces the intellectual provenance of recapitulationism and argues for its centrality to Darwin's evolutionary theory. See also Stephen Jay Gould, who provides an informative account of Haeckel and the history of the analogical or recapitulative relations between ontogeny and phylogeny in *Ontogeny and Phylogeny* (Cambridge, MA and London: Harvard University Press, 1977).

6. Robert J. C. Young, *Colonial Desire: Hybridity in Theory, Culture, and Race* (London and New York: Routledge, 1995), 35. For representative stadialist works of the Scottish Enlightenment, see Adam Ferguson, *History of Civil Society* (1767), and John Millar, *Origin of the Distinction of Ranks; Or, An Inquiry into the Circumstances Which Give Rise to Influence and Authority in the Different Members of Society* (1771).

7. See E. B. Tylor, *Primitive Culture: Researches into the Development of Mythology, Philosophy, Religion, Language, Art, and Custom* (London: John Murray, 1871); Thomas Henry Huxley, *Evidence as to Man's Place in Nature* (London and Edinburgh: Williams and Norgate, 1863). George Stocking details the genealogy of the notion of the savage in nineteenth-century Britain in *Victorian Anthropology* (New York and London: Macmillan, 1987).

8. Darwin often refers to archeological remains that provide a visual reminder of the history of so-called primitive cultures. But he contends (not at all unusually) that savages, even when literally living amongst ruins of their past, have no connection to those ruins, know nothing about what they portend – at times explicitly attributing such ignorance to a forgetting. Arguing in *The Descent of Man, and Selection in Relation to Sex* for the workings of natural selection among humans at the level of the tribe, for instance, Darwin observes: "All that we know about savages, or may infer from their traditions and from old monuments, the history of which is

quite forgotten by the present inhabitants, shew that from the remotest times successful tribes have supplanted other tribes." Charles Darwin, *The Descent of Man, and Selection in Relation to Sex*, 2 vols. (1871; reprint, Princeton, NJ: Princeton University Press, 1981), 1: 160. Subsequent citations appear in the text following the abbreviation *D.*

9. Nancy Stepan, *The Idea of Race in Science: Great Britain 1800–1960* (Hamden, CT: Archon, 1982), 50.

10. Robert FitzRoy, *Proceedings of the Second Expedition, 1831–1836, under the Command of Captain Robert Fitz-Roy* [*sic*], vol. II, *Narrative of the Surveying Voyages of H.M.S. Adventure and Beagle, between the Years 1826 and 1836, Describing Their Examination of the Southern Shores of South America, and the Beagle's Circumnavigation of the Globe*, ed. FitzRoy (London: Henry Colburn, 1839), 17.

11. D. Graham Burnett undertakes a rich and illuminating treatment of the work of mapmaking, and specifically of nineteenth-century British map-making in South America, in *Masters of All They Surveyed: Exploration, Geography, and a British El Dorado* (Chicago and London: University of Chicago Press, 2000). On the culture of investment in the Victorian era, including investment in Latin America, see Cannon Schmitt, Nancy Henry, and Anjali Arondekar, eds., "Victorian Investments," special issue, *Victorian Studies* 45.1 (2002), and Nancy Henry and Cannon Schmitt, eds., *Victorian Investments: New Perspectives on Culture and Finance* (Bloomington and Indianapolis: Indiana University Press, 2008).

12. Janet Browne, *Charles Darwin: Voyaging* (New York: Knopf, 1995), 181. In the case of territories formerly controlled by Spain, the relaxation of restrictions on trade and travel followed in the wake of independence; in the case of Brazil, the Portuguese provided access to the British as a result of closer ties between the two nations after 1808, when Britain placed the Portuguese royal family and court beyond the reach of Napoleon's troops by relocating them from Lisbon to Rio de Janeiro. See C.K. Webster, ed., *Britain and the Independence of Latin America, 1812–30: Select Documents*, 2 vols. (London and New York: Oxford University Press, 1938); G.S. Graham and R.A. Humphreys, eds., *The Navy and South America, 1807–23* (Cambridge: Cambridge University Press, 1962); William Kaufmann, *British Policy and the Independence of Latin America, 1804–1828* (New Haven, CT: Yale University Press, 1951).

13. Charles Lyell, *Principles of Geology, Being an Attempt to Explain the Former Changes of the Earth's Surface, by Reference to Causes Now in Operation*, 3 vols. (1830–33; reprint, Chicago: University of Chicago Press, 1991), III: 384. FitzRoy presented Darwin with a copy of the first volume of Lyell's *Principles* immediately before the *Beagle* set sail from England; Darwin received the two subsequent volumes while on the voyage. Browne, *Charles Darwin*, 186–89; Adrian Desmond and James Moore, *Darwin: The Life of a Tormented Evolutionist* (New York and London: Norton, 1991), 131. On "deep time," see Stephen Jay Gould, *Time's Arrow, Time's Cycle: Myth*

and Metaphor in the Discovery of Geological Time (Cambridge, MA and London: Harvard University Press, 1987); Sandra Herbert, *Charles Darwin, Geologist* (Ithaca, NY and London: Cornell University Press, 2005); Noah Heringman, *Romantic Rocks, Aesthetic Geology* (Ithaca, NY and London: Cornell University Press, 2004); Martin J. S. Rudwick, *Bursting the Limits of Time: The Reconstruction of Geohistory in the Age of Revolution* (Chicago and London: University of Chicago Press, 2005) and *Scenes from Deep Time: Early Pictorial Representations of the Prehistoric World* (Chicago and London: University of Chicago Press, 1995); and Virginia Zimmerman, *Excavating Victorians* (Albany: State University of New York Press, 2008).

14. Charles Darwin, *The Autobiography of Charles Darwin, 1809–1882* (1887; reprint, New York and London: Norton, 1969), 21.

15. *Ibid.*, 140.

16. *Ibid.*, 137–38.

17. Browne, *Charles Darwin*, 193–94; Darwin, *Autobiography*, 78. On Victorian science and empire, see Jane Camerini, "Remains of the Day: Early Victorians in the Field," in *Victorian Science in Context*, ed. Bernard Lightman (Chicago: University of Chicago Press, 1997), 354–77; Richard Drayton, *Nature's Government: Science, Imperial Britain, and the "Improvement" of the World* (New Haven, CT and London: Yale University Press, 2000); and Jim Endersby, *Imperial Nature: Joseph Hooker and the Practices of Victorian Science* (Chicago and London: University of Chicago Press, 2008). On twentieth- and twenty-first-century mass media as prosthetic memory, see Alison Landsberg, *Prosthetic Memory: The Transformation of American Remembrance in the Age of Mass Culture* (New York: Columbia University Press, 2004).

18. Darwin to Caroline Darwin, April 12, 1833:

> We here saw the native Fuegian; an untamed savage is I really think one of the most extraordinary spectacles in the world. – the difference between a domesticated & wild animal is far more strikingly marked in man. – in the naked barbarian, with his body coated with paint, with difficulty we see a fellow-creature. – No drawing or description will at all explain the extreme interest which is created by the first sight of savages. – It is an interest which almost repays one for a cruize in these latitudes; & this I assure you is saying a good deal. (Frederick Burkhardt, Sydney Smith, *et al.*, eds., *The Correspondence of Charles Darwin*, 16 vols. (Cambridge: Cambridge University Press, 1985–), II: 303)

19. Darwin, *Autobiography*, 80.

20. Charles Darwin, *Journal of Researches into the Natural History and Geology of the Various Countries Visited by H.M.S. Beagle, under the Command of Captain FitzRoy, R.N, from 1832 to 1836* (1839), vols. II and III, *The Works of Charles Darwin*, 29 vols. (London: Pickering, 1986–89), III: 473, 474. Subsequent citations appear in the text following the abbreviation *J*. Similar language occurs in the letters. See, for instance, Darwin to J. S. Henslow, April 11, 1833:

> The Fuegians are in a more miserable state of barbarism, than I had expected ever to have seen a human being … I shall never forget, when entering Good Success

Bay, the yell with which a party received us. They were seated on a rocky point, surrounded by the dark forest of beech; as they threw their arms wildly round their heads & their long hair streaming they seemed the troubled spirits of another world.

See also Darwin to Charles Whitley, July 23, 1834: "But, I have seen nothing, which more completely astonished me, than the first sight of a Savage; It was a naked Fuegian his long hair blowing about, his face besmeared with paint. There is in their countenances, an expression, which I believe to those who have not seen it, must be inconceivably wild." Burkhardt and Smith, *et al.*, *Correspondence*, 1: 306–07, 397.

21. Darwin is somewhat unusual in treating the terms "savage" and "barbarian" as synonymous. More common is John Stuart Mill's usage in an 1836 essay on "Civilization," where he asserts that civilization "stands for that kind of improvement only, which distinguishes a wealthy and powerful nation from savages or barbarians" – the "or" setting apart the two latter types of people from each other as both are set apart from the civilized. "Civilization" (1836), in *Dissertations and Discussions, Political, Philosophical, and Historical* (third edition; London: Longmans, Green, Reader, and Dyer, 1875), 1: 160. Mill employs a tripartite division common as far back as the Greeks. Relatively new, however, is his proto-evolutionary sense of savages, barbarians, and civilized peoples inhabiting different points on a developmental time-line. Stocking, *Victorian Anthropology* 8–14. The distinction as it appears in Mill, which was only strengthened in the wake of Darwinian evolutionary theory, continues active right through to the end of the nineteenth century and beyond. Witness, for instance, this sentence from the first chapter of Thorstein Veblen's *The Theory of the Leisure Class: An Economic Study in the Evolution of Institutions* (1899; reprint, New York: Modern Library, 1934): "At a step backward in the cultural scale [from "lower barbarian culture"] – among savage groups – the differentiation of employments is still less elaborate" (5–6).

22. Compare the account of first contact with Fuegians written on December 18, 1832 in the diary Darwin kept during the voyage of the *Beagle*:

It was without exception the most curious spectacle I ever beheld. I would not have believed how entire the difference between savage & civilized man is. It is greater than between a wild & domesticated animal, in as much as in man there is greater power of improvement … I believe if the world was searched, no lower grade of man could be found. (*Diary of the Voyage of H.M.S. Beagle*, vol. 1, *The Works of Charles Darwin*, 111)

23. Cook, quoted in J. C. Beaglehole, *The Life of Captain James Cook* (Stanford, CA: Stanford University Press, 1974), 161. But, as David N. Livingstone notes, the Fuegians depicted by Alexander Buchan as "squalid" and "living a life of misery" became, in engravings John Hawkesworth placed in the official account of the voyage, *An Account of the Voyages Undertaken by the Order of His Present Majesty for Making Discoveries in the Southern Hemisphere* (1773), "exemplars of austere virtuousness." *Putting Science in Its*

Place: Geographies of Scientific Knowledge (Chicago and London: University of Chicago Press, 2003), 165, 164. Gayatri Spivak points out that Kant in the *Critique of Judgment* invokes Fuegians, along with Australian aborigines, as the epitome of the not-quite or barely human: "we do not see why it is necessary that men should exist (a question which is not so easy to answer if we cast our thoughts by chance on the New Hollanders or the inhabitants of Tierra del Fuego)." *A Critique of Postcolonial Reason: Toward a History of the Vanishing Present* (Cambridge, MA and London: Harvard University Press, 1999), 26.

24. Peter Hulme, "Cast Away: The Uttermost Parts of the Earth," in *Sea Changes: Historicizing the Ocean*, ed. Bernhard Klein and Gesa Mackenthun (New York and London: Routledge, 2004), 190.

25. Watter Benn Michaels, "'You who never was there': Slavery and the New Historicism, Deconstruction and the Holocaust," *Narrative* 4 (1996): 4.

26. *Charles Darwin's Notebooks, 1836–1844: Geology, Transmutation of Species, Metaphysical Enquiries*, ed. Paul H. Barrett *et al.* (Ithaca, NY: Cornell University Press, 1987), 521. See also the following from the C notebook, kept during the first half of 1838: "Memory springing up after long intervals of forgetfulness. – after sleep, <> analogies with memory in offspring." *Ibid.*, 293. Perhaps the most resonant formulation along these lines occurs just a manuscript page later in the C notebook: "Analogy. a bird can swim without being web footed yet with much practice & led on by circumstanc [*sic*] it becomes web footed, now Man by effort of Memory can remember how to swim after having once learnt, & if that was a regular contingency the brain would become webfooted & there would be no act of memory." *Ibid.*

27. Laura Otis, *Organic Memory: History and the Body in the Late Nineteenth and Early Twentieth Centuries* (Lincoln: University of Nebraska Press, 1994).

28. For another account of the encounter between the crew of the *Beagle* and the inhabitants of Tierra del Fuego, see Michael Taussig's *Mimesis and Alterity: A Particular History of the Senses* (New York and London: Routledge, 1993). Taussig views this encounter as a signal moment in the history of the contemporary West's relation to "mimesis," here understood as "the nature that culture uses to create second nature, the faculty to copy, imitate, make models, explore difference, yield into and become Other" (xiii). See also Anne Mackaye Chapman, *Darwin in Tierra del Fuego* (Buenos Aires: Imago Mundi, 2006) and *Cape Horn: Encounters with the Native People Before and After Darwin* (forthcoming); Stephen Jay Gould, *The Mismeasure of Man* (New York: Norton, 1981); Nick Hazlewood, *Savage: The Life and Times of Jemmy Button* (London: Hodder and Stoughton, 2000); Richard Darwin Keynes, *Fossils, Finches and Fuegians: Darwin's Adventures and Discoveries on the Beagle* (Oxford and New York: Oxford University Press, 2003); Paul Magee, *From Here to Tierra del Fuego* (Urbana and Chicago: University of Illinois Press, 2000); Alan Moorehead, *Darwin and the Beagle* (London: Hamilton, 1969); Keith S. Thompson, *The Story of Darwin's Ship* (New York:

Norton, 1995); and Ian Duncan, "Darwin and the Savages," *Yale Journal of Criticism* 4 (1991): 13–45.

29. Stocking, *Victorian Anthropology*, 15. In the *Descent*, Darwin states his faith in the comparative method directly: "[W]e are chiefly concerned with primeval times, and our only means of forming a judgment on this subject is to study the habits of existing semi-civilised and savage nations." Darwin, *Descent*, II: 338. See also the companion volume to the *Descent*, *The Expression of the Emotions in Man and Animals*, in which Darwin declares that "the essence of savagery seems to consist in the retention of a primordial condition." *The Expression of the Emotions in Man and Animals* (second edition, 1890), vol. XXIII, *The Works of Charles Darwin*, 179. Thomas Henry Huxley makes a similar claim: "[T]he ethnologist may turn to the practical life of men; and relying upon the inherent conservatism and small inventiveness of untutored mankind, he may hope to discover, in manners and customs, or in weapons, dwellings, and other handiwork, a clue to the origin of the resemblances and differences of nations." T. H. Huxley, "On the Methods and Results of Ethnology" (1865), reprinted in *Critiques and Essays* (New York: Appleton, 1887), 135. See also Huxley's offhanded statement that "Half-civilized people are essentially children." T. H. Huxley, "Science at Sea," *Westminster Review* 61 (1854): 98–119.

30. Peter Mason, *The Lives of Images* (London: Reaktion, 2001), 46–47.

31. In connection with Jemmy Button, I should perhaps write "supposedly purchased." According to Lucas Bridges, son of the Reverend Thomas Bridges (Anglican missionary at Ushuaia, Tierra del Fuego, and a correspondent of Darwin's), this account is "ridiculous … as no native would have sold his child in exchange for H.M.S. *Beagle* with all it had on board." *Uttermost Part of the Earth* (New York: Dutton, 1949), 30. A fourth Fuegian, given what is in the present context the especially resonant name Boat Memory to mark the relation of his capture to the incident that instigated it, the Fuegians' suspected theft of a whale boat, died of smallpox at the Plymouth naval hospital. FitzRoy, *Proceedings of the Second Expedition*, 8–10. According to Peter Mason, Fuegia Basket's actual name was Yakcushlu, and she was of "mixed Yahgan-Alakaluf parentage"; El'leparu (renamed York Minster) and Orurdelicone (Jemmy Button) were Yahgans. *The Lives of Images*, 22.

32. Gillian Beer, *Open Fields: Science in Cultural Encounter* (Oxford: Clarendon, 1996), 67.

33. Sandra Herbert writes of Darwin's initial discussion of Fuegians:

> By citing the relationship of a wild animal to its domestic cousin, Darwin was claiming identity in an original progenitor for savage and civilized man (indeed on almost every occasion he made clear that he believed all beings of human form to be of a single species) while claiming differences of a sort which would enable various groups to be ranked on a scale ranging from savage to civilized.

> But that scale, Herbert declares, was "not transmutationist." Sandra Herbert, "The Place of Man in the Development of Darwin's Theory of

Transmutation, Part I. To July 1837," *Journal of the History of Biology* 7.2 (Fall 1974): 228, 229. Such a scale was, however, well suited to transformation into a transmutationist one, as Darwin's subsequent invocations of it show.

34. P. Parker King, *Proceedings of the First Expedition, 1826–1830, under the Command of Captain P. Parker King*, vol. 1, *Narrative of the Surveying Voyages*, ed. Robert Fitzroy (London: Henry Colburn, 1839), 24.

35. FitzRoy, *Proceedings of the Second Expedition*, 1–2.

36. Nigel Leask, "Darwin's 'Second Sun': Alexander von Humboldt and the Genesis of *The Voyage of the Beagle*," in *Literature, Science, Psychoanalysis, 1830–1970: Essays in Honor of Gillian Beer*, ed. Helen Small and Trudi Tate (Oxford and New York: Oxford University Press, 2003), 28.

37. Farce: Button, Minster, and Basket were to have formed the nucleus of a Christian mission presided over by a catechist named Richard Matthews, whom the *Beagle* left behind early in 1833; the plan collapsed almost immediately: when the *Beagle* returned a few weeks later to check on its progress, Matthews "ran towards the captain's launch screaming with terror and refused to go ashore again." Browne, *Charles Darwin: Voyaging*, 252–53. Tragedy: on November 6, 1859, Jemmy Button was involved in the massacre of another group of missionaries to Tierra del Fuego, seven Britons and one Swede, at Wulaia Cove. Hazlewood, *Savage*, 209.

38. Browne, *Charles Darwin: Voyaging*, 246.

39. Stepan notes of nineteenth-century race scientists that "their deepest commitment seems to have been to the notion that the social and cultural differences observed between peoples should be understood as realities of nature." *Idea of Race*, xx.

40. FitzRoy, *Proceedings of the Second Expedition*, 2.

41. Jonathan Lamb provides a galvanizing treatment of the European encounter with Polynesia in *Preserving the Self in the South Seas, 1680–1840* (Chicago and London: University of Chicago Press, 2001).

42. Marianna Torgovnick, *Gone Primitive: Savage Intellects, Modern Lives* (Chicago and London: University of Chicago Press, 1990), 3.

43. Charles Darwin, *On the Origin of Species by Means of Natural Selection, or the Preservation of Favoured Races in the Struggle for Life* (1859; reprint, Cambridge, MA and London: Harvard University Press, 1964), 488. Subsequent citations appear in the text following the abbreviation *O*.

44. Compare a similar sentence in the conclusion to the *Descent*: "He who is not content to look, like a savage, at the phenomena of nature as disconnected, cannot any longer believe that man is the work of a separate act of creation" (*D* II: 386). Darwin employs a more elaborately developed simile in the next-to-last paragraph of *The Variation of Plants and Animals under Domestication*:

> If it were explained to a savage utterly ignorant of the art of building, how the edifice [a building constructed from "fragments at the base of a precipice"] had been raised stone upon stone, and why wedge-formed fragments were used for the arches, flat stones for the roof, etc.; and if the use of each part and of the whole building were

pointed out, it would be unreasonable if he declared that nothing had been made clear to him, because the precise cause of the shape of each fragment could not be told. But this is a nearly parallel case with the objection that selection explains nothing, because we know not the cause of each individual difference in the structure of each being. (*The Variation of Plants and Animals under Domestication* (1868), vol. xx, *The Works of Charles Darwin*, 370–71)

In his own comments on savages' response to "grand or complicated object[s]," FitzRoy accounts for the astonishment of Fuegians viewing a ship by likening it to the reaction of one of his English readers standing near a steam-powered locomotive for the first time. *Proceedings of the Second Expedition*, 3.

45. For an account of *The Descent* as a treatise on race given urgency and importance by Darwin's abolitionism, see James Moore and Adrian Desmond, "Introduction," in *The Descent of Man, and Selection in Relation to Sex*, by Charles Darwin (London: Penguin, 2004), xi–lviii.

46. Thus Gillian Beer on Darwin and Fuegians: "Darwin communicates a sense of fascinated helplessness at finding himself unable to interpret the profound difference of the other man … Here no *relation*, in the sense of message or narrative, can be established. The other is 'inconceivably wild.' But that which is inconceivable is also here a mirror-image" (*Open Fields*, 25).

47. Darwin to Kingsley, February 6, 1862, in Burkhardt and Smith, *et al.*, *Correspondence*, x: 71. See also Neville Hoad, "Wild(e) Men and Savages: The Homosexual and the Primitive in Darwin, Wilde and Freud" (Ph.D. diss., Columbia University, 1998), 30–31.

48. See especially the following paragraph:

It is almost superfluous to state that animals have excellent *Memories* for persons and places. A baboon at the Cape of Good Hope, as I have been informed by Sir Andrew Smith, recognized him with joy after an absence of nine months. I had a dog who was savage and averse to all strangers, and I purposely tried his memory after an absence of five years and two days. I went near the stable where he lived, and shouted to him in my old manner; he showed no joy, but instantly followed me out walking and obeyed me, exactly as if I had parted with him only half-an-hour before. A train of old associations, dormant during five years, had thus instantly been awakened in his mind. Even ants, as P. Huber has clearly shewn, recognised their fellow-ants belonging to the same community after a separation of four months. Animals can certainly by some means judge of the intervals of time between recurrent events. (*D* i: 45)

Darwin's commitment to establishing the fundamental similarity between the mental faculties of animals and humans persists right down to the final book he published during his lifetime, *The Formation of Vegetable Mould, Through the Action of Worms, with Observations on Their Habits* (1881; reprint, Chicago and London: University of Chicago Press, 1985), half a chapter of which is devoted to demonstrating that "worms, although standing low in the scale of organization, possess some degree of intelligence" (98). As George Levine maintains,

anthropomorphism is not usually for Darwin a mere sentimental lapse. It is rather a quite seriously worked out way of regarding a world in which there is an absolute continuity between humans and other animals. It is not so much anthropomorphism, then, as zoomorphism: that is, humans are animals, and therefore one can – as an animal oneself – understand non-human behavior simply by imagining one's way into the animal's mind. (*Darwin Loves You: Natural Selection and the Re-enchantment of the World* (Princeton, NJ: Princeton University Press, 2006), 197)

And so Darwin's insistence on the human capacities of animals necessitates a concomitant bestialization of humans, and of savages in particular; as John Durant pithily observes: "Thus was anthropomorphic zoology combined with zoomorphic anthropology in effecting the unification of animals and man, matter and mind, nature and morality." "The Ascent of Nature in Darwin's *Descent of Man*," in *The Darwinian Heritage*, ed. David Kohn (Princeton, NJ: Princeton University Press, 1985), 292.

49. Hulme, "Cast Away," 191.
50. Hudson, *Idle Days*, 223, 226.
51. *Ibid.*, 205.
52. Torgovnick, *Gone Primitive*, 185.
53. Hudson, *Idle Days*, 214.
54. Darwin, *Autobiography*, 79. Compare the following from *Expression of the Emotions*: "I remember once seeing a boy who had just shot his first snipe on the wing, and his hands trembled to such a degree from delight, that he could not for some time reload his gun; and I have heard of an exactly similar case with an Australian savage, to whom a gun had been lent" (51). In a footnote, Francis Darwin identifies the boy as Charles Darwin himself (51n).
55. As Robert J. Richards avers:

In finding the antecedents of human rationality in animal sources, Darwin really opened no new epistemological ground … But [for pre-Darwinian thinkers] no animal … gave any hint of what was truly distinctive of human mind – namely, moral judgement. If Darwin were to solidify his case for the descent of man from lower animals, he would have to discover the roots of moral behaviour even among those creatures. And so he did. ("Darwin on Mind, Morals and Emotions," in *The Cambridge Companion to Darwin*, ed. Jonathan Hodge and Gregory Radick (Cambridge and New York: Cambridge University Press, 2003), 95–96)

56. That "the hard-worked wife of a degraded Australian savage" stands as exemplary of savagery in this instance reveals the necessity for an additional analysis, beyond the scope of this chapter, attentive to Darwin's understanding of the "savage" inability to "exert … self-consciousness" as at least in part the result of a particular set of gendered social arrangements.
57. Comparisons between the earth's geology and a book either imperfectly preserved or difficult to interpret appear throughout; see, for instance, the opening of the third volume: "As the first theorists possessed but scanty acquaintance with the present economy of the animate and inanimate world … we find them in the situation of novices, who attempt to read a history written in a foreign language." Lyell, *Principles of Geology*, III: 1.
58. Gillian Beer, *Darwin's Plots: Evolutionary Narrative in Darwin, George Eliot and Nineteenth-Century Fiction* (1983; revised edition, Cambridge and New York: Cambridge University Press, 2000), 36.

59. Adam Phillips, *Darwin's Worms: On Life Stories and Death Stories* (New York: Basic Books, 2000), 126.

60. As Phillip Barrish writes: "Darwin … recognizes that variation makes the existence of definable species, strictly speaking, a delusion, an arbitrary imposition by the naturalist." "Accumulating Variation: Darwin's *On the Origin of Species* and Contemporary Literary and Cultural Theory," *Victorian Studies* 34.3 (Summer 1991): 436. Darwin's contemporaries immediately understood this implication of his work. In *Studies in Animal Life* (London: Smith, Elder, 1862), for instance, George Henry Lewes exorts: "Let us never forget that Species have no existence. Only individuals exist, and these all vary more or less from one another" (155). For an incisive consideration of the current debate on the "species barrier," see Barbara Herrnstein Smith, "Animal Relatives, Difficult Relations," *differences* 15.1 (2004): 1–19.

61. "Closeness, meaning at once kinship and resemblance, is the ground of conflict." Duncan, "Darwin and the Savages," 27.

2 ALFRED RUSSEL WALLACE'S TROPICAL MEMORABILIA

1. Alfred Russel Wallace, *A Narrative of Travels on the Amazon and Rio Negro, with an Account of the Native Tribes, and Observations on the Climate, Geology, and Natural History of the Amazon Valley* (London: Reeve and Co., 1853), iv. Subsequent citations appear in the text following the abbreviation *N*.

2. More precisely, Wallace was *relatively* safe aboard the *Jordeson*, a vessel in a poor state of repair – "she leaked tremendously," Wallace observes – and, by the end of the voyage it made from Cuba to England, very short on rations (*N* 402).

3. The loss of Wallace's collections meant losing his memories but also losing money and, with it, the ability to continue to pursue a career in natural history. Lacking the family fortune possessed by Darwin and similar gentleman naturalists, Wallace had to fund his own travels. He did so with the sale of specimens. Collecting was a business venture, and it was only the decision of his agent, Samuel Stevens, to insure Wallace's South American collections that prevented their loss from being a financial as well as mnemonic catastrophe. For some of the details of Wallace's arrangements on this front as well as a wider discussion of natural history collecting and fieldwork in the nineteenth century, see Jane Camerini, "Wallace in the Field," *Osiris*, second series, 11 (1996): 44–65; Camerini, "Remains of the Day: Early Victorians in the Field," in *Victorian Science in Context*, ed. Bernard Lightman (Chicago: University of Chicago Press, 1997), 354–77; and Nancy Leys Stepan, *Picturing Tropical Nature* (Ithaca, NY: Cornell University Press, 2001), 31–32, 57–84.

4. For earlier versions of Wallace's complaint about civilization's ruination of South American indigenes, see D. Graham Burnett, "'It Is Impossible to Make a Step without the Indians': Nineteenth-Century Geographical

Exploration and the Amerindians of British Guiana," *Ethnohistory* 49.1 (Winter 2002): 3–40.

5. Alfred Russel Wallace, *My Life: A Record of Events and Opinions*, 2 vols. (London: Chapman and Hall, 1905), 1: 257. Subsequent citations appear in the text following the abbreviation *ML*.

6. Wallace defended *Vestiges* against Bates's dismissal (*ML* 1: 255). James A. Secord notes of the effect of the book on Wallace: "He had a commitment to socialism, self-help, and phrenology. But it was partly through *Vestiges* that he had found his vocation as a naturalist and also his problem, formulating 'the law which has regulated the introduction of new species.'" *Victorian Sensation: The Extraordinary Publication, Reception, and Secret Authorship of Vestiges of the Natural History of Creation* (Chicago and London: University of Chicago Press, 2000), 333. Of Wallace's search for evidence in connection with the origin of species, Martin Fichman writes: "The youthful surveyor and amateur natural historian was no naïve Baconian. He already had a theoretical framework that his future empirical observations and field-work would either strengthen or render defective." *An Elusive Victorian: The Evolution of Alfred Russel Wallace* (Chicago and London: University of Chicago Press, 2004), 69.

7. On receiving Wallace's paper, Darwin wrote to Charles Lyell on June 18, 1858: "Your words have come true with a vengeance – that I should be forestalled … I never saw a more striking coincidence; if Wallace had my MS. sketch written out in 1842, he could not have made a better short abstract!" *The Life and Letters of Charles Darwin, Including an Autobiographical Chapter*, ed. Francis Darwin, 3 vols. (London: John Murray, 1887), 11: 116. The paper, which argues that the struggle for existence among species resulted in continual divergence eventually amounting to speciation, was presented to the Linnean Society on July 1, 1858, together with two pieces by Darwin: an abstract of a letter he had sent to Asa Gray in 1857 and an extract from "an unpublished Work on Species." Wallace's paper and Darwin's abstract and extract were then published together, following a prefatory letter from Charles Lyell and Joseph Hooker; see the *Journal of the Proceedings of the Linnean Society* 3 (1858): 45–62. On the entire episode as well as its role in the history of evolutionary theory, see Barbara G. Beddall, "Wallace, Darwin, and the Theory of Natural Selection: A Study in the Development of Ideas and Attitudes," *Journal of the History of Biology* 5.1 (1968): 261–323; Arnold C. Brackman, *A Delicate Arrangement: The Strange Case of Charles Darwin and Alfred Russel Wallace* (New York: Columbia University Press, 1980); John Langdon Brooks, *Just before the Origin: Alfred Russel Wallace's Theory of Evolution* (New York: Columbia University Press, 1984); Janet Browne, *Charles Darwin: The Power of Place* (Princeton, NJ and Oxford: Princeton University Press, 2002); Charles Darwin, "Introduction," *On the Origin of Species by Means of Natural Selection, or the Preservation of Favoured Races in the Struggle for Life* (1859; reprint, Cambridge, MA and London: Harvard University Press, 1964), 1–6; Martin Fichman, *Evolutionary Theory and Victorian Culture* (Amherst, NY:

Humanity, 2002); Wilma B. George, *Biologist Philosopher: A Study of the Life and Writings of Alfred Russel Wallace* (London and New York: Abelard-Schuman, 1964); Bert James Loewenberg, *Darwin, Wallace and the Theory of Natural Selection; Including the Linnean Society Papers* (New Haven, CT: G. E. Cinamon, 1957); James Moore, "Wallace's Malthusian Moment: The Common Context Revisited," in *Victorian Science in Context*, ed. Bernard Lightman (Chicago: University of Chicago Press, 1997), 290–311; H. Lewis McKinney, *Wallace and Natural Selection* (New Haven, CT and London: Yale University Press, 1972); Michael Shermer, *In Darwin's Shadow: The Life and Science of Alfred Russel Wallace* (Oxford and New York: Touchstone, 2002); Ross A. Slotten, *The Heretic in Darwin's Court: The Life of Alfred Russel Wallace* (New York: Columbia University Press, 2004); and Robert M. Young, *Darwin's Metaphor: Nature's Place in Victorian Culture* (Cambridge: Cambridge University Press, 1985).

8. Alfred Russel Wallace, "On the Tendency of Varieties to depart indefinitely from the Original Type," *Journal of the Proceedings of the Linnean Society* 3 (1858), 59; emphasis in the original here and throughout unless otherwise noted. John Durant writes: "From the outset, he [Wallace] had one eye on the problem of human origins." "Scientific Naturalism and Social Reform in the Thought of Alfred Russel Wallace," *The British Journal for the History of Science* 12 (1979), 39. Michael Shermer concurs: "There is little doubt that Wallace went to both the Amazon and the Malay Archipelago, at least in large part, to understand the origins of humanity and society." *In Darwin's Shadow*, 236–37. On the place of ethnography in Wallace's and Darwin's speculations on the origin of species, see also McKinney, *Wallace and Natural Selection*, 82–96.

9. Fichman, *An Elusive Victorian*, 68.

10. Alfred Russel Wallace, "The Origin of Human Races and the Antiquity of Man Deduced from the Theory of Natural Selection," *Journal of the Anthropological Society of London* 2 (1864): clvii-clxxxvii. This article, which Durant calls "the first important application of the theory of natural selection to mankind," and which Janet Browne adduces when she calls Wallace "the first of the Darwinians deliberately to apply natural selection to the emergence of human difference," was originally delivered as a lecture to the Anthropological Society of London on March 1, 1864. Durant, "Scientific Naturalism," 32; Browne, *Charles Darwin: The Power of Place*, 253.

11. Alfred Russel Wallace, "Sir Charles Lyell on Geological Climates and the Origin of Species," *Quarterly Review* 126 (April 1869): 391–92. Subsequent citations appear in the text following the abbreviation "Lyell." Robert M. Young glosses these claims with the explanation that "Wallace's partial defection can best be described as a loss of faith in the plausibility of the principle of utility as an adequate way of accounting for the development of certain key structures in man." *Darwin's Metaphor*, 113.

12. There were other factors, including his impoverished upbringing, the time he spent as a surveyor involved in land enclosure in Wales, his informal

education among working-class intellectuals at Mechanics' Institutes, and his study of socialist political theory. See Durant, "Scientific Naturalism"; Fichman, *An Elusive Victorian*. In a quantitative approach to history and biography, Michael Shermer determines that "anthropological experiences" were among the five most important "external forces" contributing to Wallace's dissent from Darwin's conclusions about human evolution; he identifies in addition five important "internal forces." For a graphic representation of his estimate of the relative weight of each such "force," see the "Historical Matrix Model." Given the inevitable dependence of these quantitative assessments on logically prior acts of interpretation and selection on the part of the biographer (consider, for instance, the neatly symmetrical and therefore clearly arbitrary division of forces at work into two sets of five), the reliability of the specific conclusions drawn seems doubtful. More convincing is Shermer's less numerically oriented account of Wallace as a "heretic personality." *In Darwin's Shadow*, 8, 250–70.

13. When measured against some of the savage societies in which he has lived, Wallace contends, "our system of government, of administering justice, of national education, and our whole social and moral organization, remains in a state of barbarism." Alfred Russell Wallace, *The Malay Archipelago: The Land of the Orang-Utan, and the Bird of Paradise; A Narrative of Travel, with Studies of Man and Nature*, 2 vols. (London: Macmillan and Co., 1869), II: 461–62. Subsequent citations appear in the text following the abbreviation *MA*.

14. For a fine discussion of the prehistory of "salvage anthropology" and related forms of exoticism, see Chris Bongie, *Exotic Memories: Literature, Colonialism, and the Fin de Siècle* (Stanford, CA: Stanford University Press, 1991).

15. Robert J. Richards conducts an illuminating investigation of the place of teleological and nonteleological aspects of Darwin's theory in *The Meaning of Evolution: The Morphological Construction and Ideological Reconstruction of Darwin's Theory* (Chicago and London: University of Chicago Press, 1992).

16. Charles Darwin, *The Autobiography of Charles Darwin, 1809–1882* (1887; reprint, New York and London: Norton, 1969), 21.

17. *Ibid.*, 80.

18. On nakedness, "savages," and imperialism, see Philippa Levine, "States of Undress: Nakedness and the Colonial Imagination," *Victorian Studies* 50.2 (Winter 2008): 189–219.

19. Wallace also repeats Darwin's comparison of "savages" to wild animals and "civilized" peoples to domesticated ones; see Charles Darwin, *Journal of Researches into the Natural History and Geology of the Various Countries Visited by H.M.S. Beagle, under the Command of Captain FitzRoy, R.N, from 1832 to 1836* (1839), vols. II and III, *The Works of Charles Darwin*, 29 vols. (London: Pickering, 1986–89), II: 178.

20. The other is Alexander von Humboldt's *Personal Narrative of Travels to the Equinoctial Regions of the New Continent, During the Years 1799–1804*, translated into English by Helen Maria Williams between 1814 and 1829. In *A*

Narrative of Travels, Wallace credits William Edwards's *A Voyage up the Amazon, Including a Residence at Pará* (1847) with his decision to explore specifically the Amazon region "on account of its easiness of access and the little that was known of it compared with most other parts of South America" (*N* iii).

21. Darwin, *Journal of Researches*, 474.

22. The companion, we learn in *A Narrative of Travels*, was Senhor João Antonio de Lima, "a middle-sized, wiry, grizzly man, with a face something like the banished lord in the National Gallery" (*N* 134). The "banished lord" refers to a painting by Sir Joshua Reynolds.

23. I discuss the affective registers of natural-historical collecting at more length in "Victorian Beetlemania," in *Victorian Animal Dreams: Representations of Animals in Victorian Literature and Culture*, ed. Deborah Morse and Martin Danahay (Aldershot, Hampshire and Burlington, VT: Ashgate, 2007), 35–51; see also Kate Flint, "Sensuous Knowledge," in *Unmapped Countries: Biological Visions in Nineteenth Century Literature and Culture*, ed. Anne-Julia Zwierlein (London: Anthem, 2005), 207–15.

24. In this Wallace resembles P. Parker King; see his *Proceedings of the First Expedition, 1826–1830, under the Command of Captain P. Parker King*, vol. 1 of *Narrative of the Surveying Voyages of H.M.S. Adventure and Beagle, between the Years 1826 and 1836, Describing Their Examination of the Southern Shores of South America, and the Beagle's Circumnavigation of the Globe*, ed. Robert FitzRoy (London: Henry Colburn, 1839), 24.

25. For the "denial of coevalness," see Johannes Fabian, *Time and the Other: How Anthropology Makes Its Object* (New York: Columbia University Press, 1983), 31. On the comparative method, see George Stocking, *Victorian Anthropology* (New York and London: Free Press, 1987) and *Race, Culture, and Evolution: Essays in the History of Anthropology* (Chicago: University of Chicago Press, 1968).

26. But see Donna Haraway, *When Species Meet* (Minneapolis and London: University of Minnesota Press, 2007).

27. As James Moore and Adrian Desmond observe, "Unlike Wallace, who had actually lived with natives and held them in high regard, Darwin deplored the 'low morality' of peoples he never really knew." "Introduction," *The Descent of Man, and Selection in Relation to Sex*, by Charles Darwin (London: Penguin, 2004), xxxix. Compare C. Napier Bell who, in the ante-penultimate paragraph of his recollections of the years he spent among the Miskito Indians of Nicaragua, writes:

> Of course I know that it is impossible for a civilized and Christian people to look upon any savages and heathens without aversion and contempt, and regarding them, as we do the animals, "as the beasts that perish." Nevertheless, one has only to learn their language and live among them as one of themselves, to find that they have all the faculties, affections, emotions, loves, hates, and fears, just the same as ourselves, to which their innate simplicity and unsophistication lend a peculiar charm. (*Tangweera: Life and Adventures among Gentle Savages* (1899; reprint, Austin: University of Texas Press, 1989), 308)

28. Wallace's depiction of "patriarchal harmony" becomes still more striking when read alongside Darwin's depiction of what might be called maternal cacophony: "These [Fuegians] were the most abject and miserable creatures I any where beheld … [A] woman, who was suckling a recently-born child, came one day alongside the vessel, and remained there whilst the sleet fell and thawed on her naked bosom, and on the skin of her naked child." Darwin, *Journal of Researches*, 184.

29. A few pages further on he again mentions "the accounts of picture-drawing travellers, who, by only describing the beautiful, the picturesque, and the magnificent, would almost lead a person to believe that nothing of a different character could exist under a tropical sun" (*N* 9).

30. Compare Wallace on Pará:

> The general impression of the city to a person fresh from England is not very favourable. There is such a want of neatness and order, such an appearance of neglect and decay, such evidences of apathy and indolence, as to be at first absolutely painful. But this soon wears off, and some of these peculiarities are seen to be dependent on the climate. The large and lofty rooms, with boarded floors and scanty furniture, and with half-a-dozen doors and windows in each, look at first comfortless, but are nevertheless exactly adapted to a tropical country, in which a carpeted, curtained, and cushioned room would be unbearable. (*N* 7–8)

31. Stepan, *Picturing Tropical Nature*, 59.

32. *Ibid.*, 60, 61. Wallace's contemporaries perceived this as well. Everard F. im Thurn notes: "In correction of the false views [about the tropics] thus spread, Mr. Wallace's careful analysis of tropical scenery in general, in his admirable essay on Tropical Nature, is of great value." *Among the Indians of Guiana, being Sketches Chiefly Anthropologic from the Interior of British Guiana* (London: Kegan, Paul, Trench and Co., 1883), 88.

33. Stepan, *Picturing Tropical Nature*, 61.

34. Charles Darwin to Henry Walter Bates, December 3, 1861. *The Life and Letters of Charles Darwin*, II: 380.

35. Wallace quotes from a letter he wrote on October 5, 1852 to botanist and South American traveler Richard Spruce; the letter continues: "But good resolutions soon fade, and I am already only doubtful whether the Andes or the Philippines are to be the scene of my next wanderings" (*ML* I: 309). That Wallace contemplates specifically a return to South America confirms, in its way, the compensatory nature of his Malay voyage.

36. For nineteenth-century natural historians, New Guinea frequently took on the aspect of an alluring will-o'-the-wisp. In a diary entry for August 31, 1849, for instance, T. H. Huxley expressed feelings of dismay similar to Wallace's at being unable to explore this *terra incognita*, placing the blame not on his own incapacity but that of the captain of H. M. S. *Rattlesnake*:

> The mainland [of New Guinea] was not half a dozen miles off and there appeared to be some promise of a large river. Not a boat was sent to explore the coast. For anything we can say to the contrary, the land we saw may have been a succession of islands. If this is surveying, is this is the process of English Discovery, God defend

me from any such elaborate waste of time and opportunity. (Thomas Henry Huxley, *T. H. Huxley's Diary of the Voyage of H. M. S. Rattlesnake*, ed. Julian Huxley (Garden City, NY: Doubleday, 1931), 181)

37. Stepan, however, argues that *The Malay Archipelago* differs from *A Narrative of Travels* insofar as, in the later book, "Wallace had indeed learned to describe the tropics well" – that is, more in accord with convention and popular taste. *Picturing Tropical Nature*, 68.

38. Robert Aguirre discusses the politics of panoramas and similar nineteenth-century forms of representation and consumption of the Americas in *Informal Empire: Mexico and Central America in Victorian Culture* (Minneapolis and London: University of Minnesota Press, 2005).

39. David Amigoni, placing Wallace's writings in the tradition of defamiliarization, observes that "Wallace's estrangement was … grounded in a complex poet-ics of symbolic material … one that sought to re-think social relations on the basis of the strangest that has been thought and said." *Colonies, Cults, and Evolution: Literature, Science and Culture in Nineteenth-Century Writing* (Cambridge and New York: Cambridge University Press, 2007), 129.

40. Compare Darwin in The *Journal of Researches*: "A bad earthquake at once destroys the oldest associations: the world, the very emblem of all that is solid, has moved beneath our feet like a crust over a fluid; one second of time has conveyed to the mind a strange idea of insecurity, which hours of reflection would never have created." *Journal of Researches*, 289.

41. Fichman observes: "The subtitle [of *The Malay Archipelago*] and the text itself indicate that Wallace was already pondering what he would later term the question of *Man's Place in the Universe*." Fichman, *An Elusive Victorian*, 35.

42. See especially Alfred Russel Wallace, "The Native Problem in South Africa and Elsewhere," *Independent Review* (November 1906): 174–82.

43. No doubt Janet Browne has in mind passages such as these when she writes that "While Wallace, like Darwin, was for the most part a humane and cultivated man, he nonetheless endorsed Western cultural superiority and matched it to evolutionary theory." Browne, *Charles Darwin: The Power of Place*, 254. But this depiction of Wallace is partial insofar as it elides the critique of nineteenth-century European civilization made possible by his encounters with and memories of "savages."

44. See Gayatri Chakravorty Spivak's analysis of such a duty as the "terrorism of the categorical imperative." *A Critique of Postcolonial Reason: Toward a History of the Vanishing Present* (Cambridge, MA and London: Harvard University Press, 1999).

45. But compare: "Here we have a picture of true savage life; of small isolated communities at war with all around them, subject to the wants and miser-ies of such a condition, drawing a precarious existence from the luxuriant soil, and living on from generation to generation, with no desire for physical amelioration, and no prospect of moral advancement" (*MA* 381).

46. Quoted in Shermer, *In Darwin's Shadow*, 235–36. Wallace's Malay journals are now in possession of the Linnean Society of London.

47. For a compact example of such doubleness, see Wallace's commentary on the villagers in the Matabello Islands:

> The people are wretched ugly dirty savages … What a contrast between these people and such savages as the best tribes of hill Dyaks in Borneo, or the Indians of the Uaupes in South America, living on the banks of clear streams, clean in their persons and their houses, with abundance of wholesome food, and exhibiting its effect in healthy skins and beauty of form and feature! There is in fact almost as much difference between the various races of savage as of civilized peoples, and we may safely affirm that the better specimens of the former are much superior to the lower examples of the latter class. (*MA* II: 102)

48. As Jane R. Camerini points out, Thomas Henry Huxley coined the term "Wallace's Line." "Evolution, Biogeography, and Maps: An Early History of Wallace's Line," *Isis* 84.4 (December 1993), 700n. See also Fichman, *An Elusive Victorian*, 46, 96. David N. Livingstone notes that "[f]or Wallace, human geography and animal geography were always intimately intertwined. And maps became a strategic rhetorical device through which he could conjure into view both the social and the zoological facts his theories sought to explain." *Putting Science in Its Place: Geographies of Scientific Knowledge* (Chicago and London: University of Chicago Press, 2003), 162.

49. The first act in which the state really comes forward as the representative of the whole of society – the taking possession of the means of production in the name of society – is at the same time its last independent act as a state. The interference of the state power in social relations becomes superfluous in one sphere after another, and then dies away of itself. The government of persons is replaced by the administration of things and the direction of the processes of production. The state is not "abolished," it withers away. (Friedrich Engels, *Anti-Dühring* (1877; reprint, Beijing: Foreign Languages Press, 1976), 363)

On Wallace and Marxism, see D. A. Stack, "The First Darwinian Left: Radical and Socialist Responses to Darwin, 1859–1914," *History of Political Thought* 21.4 (Winter 2000): 682–710.

50. Wallace anticipates these comments while admiring the behavior of his shipmates aboard a prau: "Considering we have fifty men of several tribes and tongues on board, wild, half-savage looking fellows, and few of them feeling any of the restraints of morality or education, we get on wonderfully well. There is no fighting or quarrelling, as there would certainly be among the same number of Europeans with as little restraint upon their actions" (*MA* II: 168–69).

51. As Peter Raby argues, the conclusion to *The Malay Archipelago* "stemmed from two major thrusts of Wallace's observations and instincts: first, the surprisingly advanced nature of so-called 'savage' man and 'savage' societies; secondly, his separation of the intellectual and, especially, the moral attributes of man from the physical." Peter Raby, *Alfred Russel Wallace: A Life* (Princeton, NJ: Princeton University Press, 2001), 201–02.

52. In a bracing "Note" following the last words of the text proper ("I now bid my readers – Farewell!"), Wallace defends his use of the term "barbarism" to describe the state of nineteenth-century European civilization:

    Those who believe that our social condition approaches perfection, will think the above word [i.e., barbarism] harsh and exaggerated, but it seems to me the only word that can be truly applied to us … We allow over a hundred thousand persons known to have no means of subsistence but by crime, to remain at large and prey upon the community, and many thousand children to grow up before our eyes in ignorance and vice, to supply trained criminals for the next generation. This, in a country which boasts of its rapid increase in wealth, of its enormous commerce and gigantic manufactures, of its mechanical skill and scientific knowledge, of its high civilization and its pure Christianity, – I can but term a state of social barbarism. We also boast of our love of justice, and that the law protects rich and poor alike, yet we retain money fines as a punishment, and make the very first steps to obtain justice a matter of expense – in both cases a barbarous injustice, or denial of justice to the poor … We permit absolute possession of the soil of our country, with no legal rights of existence on the soil, to the vast majority who do not possess it. A great landholder may legally convert his whole property into a forest or a hunting-ground, and expel every human being who has hitherto lived upon it … [T]his is a power of legally destroying his fellow creatures; and that such a power should exist, and be exercised by individuals, in however small a degree, indicates that, as regards true social science, we are still in a state of barbarism. (*MA* ii: 463–64)

53. This is, in essence, a species of degenerationism – a more explicit version of which appears in Wallace's review of E. B. Tylor's *Primitive Culture* (1871):

    Another important question treated very fully is that of development and progress, which are held to be fully established, degeneration being rare and exceptional; and the passage in Sir Charles Lyell's *Antiquity of Man*, in which he sarcastically maintains that if man has degenerated we ought to find, instead of rude implements of flint and bone, lines of buried railroads and electric telegraphs, with astronomical instruments and microscopes better than any we possess, is quoted with approval. But surely this passage is illogical; for man might slowly degenerate in mind while still progressing in arts, and even in science, because these are necessarily growths, and the adapter and improver may have less genius than the inventor who went before him. Mr. Galton has carefully discussed one phase of this question in his *Hereditary Genius*, and gives good reasons for believing that the average Greek of antiquity was higher mentally than the average European of to-day; and the fact that the Greeks had neither microscopes nor even the printing machine has really no bearing whatever on the question. The conception that the human race, as a whole, was higher morally and intellectually ten thousand years ago than it is now, is not disproved by evidence of any amount of inferiority in the arts, which of course is overwhelming. (Alfred Russel Wallace, "Tylor's 'Primitive Culture,'" *The Academy* (February 15, 1872), 70)

54. Kuhn defines "normal science" as "research firmly based upon one or more past scientific achievements, achievements that some particular scientific community acknowledges for a time as supplying the foundation for its further practice." Thomas S. Kuhn, *The Structure of Scientific Revolutions* (1962; third edition, Chicago and London: University of Chicago Press, 1996), 10.

55. Darwin to Charles Lyell, May 4, 1869. *Life and Letters of Charles Darwin*,
    III: 117.

56. See *A Narrative of Travels*:

    Their figures are generally superb; and I have never felt so much pleasure in gazing
    at the finest statue, as at these living illustrations of the beauty of the human form.
    The development of the chest is such as I believe never exists in the best-formed
    European, exhibiting a splendid series of convex undulations, without a hollow in
    any part of it. (478)

    See also Wallace's Malay journal for April 6, 1857:

    Here [in the Aru Islands], as among the Dyaks of Borneo & the Indians of the
    Upper Amazon, I am delighted with the beauty of the human form, a beauty of
    which stay at home civilized people can never have any conception. What are the
    finest Grecian statues to the living moving breathing forms which every where sur-
    round me. The unrestrained grace of the naked savage as he moves about his daily
    occupations or lounges at his ease must be seen to be understood. A young sav-
    age bending his bow is the perfection of physical beauty. (Quoted in Shermer, *In
    Darwin's Shadow*, 237)

57. Indeed, this claim and the related claim it allows about prehistoric humans
    undergird the conclusions Wallace reaches in his earlier essay, "The Origin
    of Human Races and the Antiquity of Man Deduced from the Theory of
    Natural Selection":

    From the time … when the social and sympathetic feelings came into active opera-
    tion, and the intellectual and moral faculties became fairly developed, man would
    cease to be influenced by 'natural selection' in his physical form and structure; as an
    animal he would remain almost stationary; the changes of the surrounding universe
    would cease to have upon him that powerful modifying effect which it exercises
    over other parts of the organic world. But from the moment that his body became
    stationary, his mind would become subject to those very influences from which his
    body had escaped. (Wallace, "Origin of Human Races," clxiii–clxiv)

    Here already the outlines of what will come to be Wallace's exceptionalism
    in connection with humans may be seen:

    He is, indeed, a being apart, since he is not influenced by the great laws which irre-
    sistibly modify all other organic beings. Nay more; this victory which he has gained
    for himself gives him a directing influence over other existences. Man has not only
    escaped "natural selection" himself, but he actually is able to take away some of that
    power from nature which, before his appearance, she universally exercised. We can
    anticipate the time when the earth will produce only cultivated plants and domes-
    tic animals; when man's selection shall have supplanted "natural selection"; and
    when the ocean will be the only domain in which that power can be exerted, which
    for countless cycles of ages ruled supreme over all the earth. (Wallace, "Origin of
    Human Races," cxviii)

58. Huxley asserts that "the matter of life [is] composed of ordinary matter, dif-
    fering from it only in the manner in which its atoms are aggregated." "On
    the Physical Basis of Life" (1868), reprinted in T. H. Huxley, *Collected Essays*,
    9 vols. (New York: Appleton, 1894–98), I: 145.

59. Alfred Russel Wallace, "The Limits of Natural Selection as Applied to Man," in *Contributions to the Theory of Natural Selection: A Series of Essays* (London: Macmillan and Co., 1870), 335–36. Subsequent citations appear in the text following the abbreviation "Limits."

60. Compare the appendix to *The Malay Archipelago*, "On the Crania and the Languages of the Races of Man in the Malay Archipelago": on one hand, Wallace states unequivocally that cranial measurements are "of very little value" in classifying humans; on the other, he enlists craniometry in support of his claim that "the Malays and Papuans are radically distinct races" (*MA* II: 467, 470).

61. Charles Darwin, *The Descent of Man, and Selection in Relation to Sex*, 2 vols. (1871; reprint, Princeton, NJ: Princeton University Press, 1981), II: 404.

62. Sigmund Freud, "A Difficulty in the Path of Psychoanalysis" (1917), reprinted in *The Standard Edition of the Complete Psychological Works of Sigmund Freud*, trans. James Strachey *et al.*, 24 vols. (London: Hogarth Press, 1955–74), XVII: 136–44. Gillian Beer discusses Darwinism as the second of Freud's "three blows" in *Darwin's Plots: Evolutionary Narrative in Darwin, George Eliot and Nineteenth-Century Fiction* (1983; revised edition, Cambridge and New York: Cambridge University Press, 2000), 8–14.

63. Wallace, "The Native Problem," 182. See also the much earlier comments along these lines made in his review of Tylor's *Primitive Culture*: "One of the most important results of Mr. Tylor's researches, and that which is most clearly brought out in every part of his work, is, that for the purpose of investigating the development of man's mental nature race may be left out of the question, and all mankind treated as essentially one." Wallace, "Tylor's 'Primitive Culture,'" 70.

3 CHARLES KINGSLEY'S RECOLLECTED EMPIRE

1. The formulation belongs to Victor Wolfgang Von Hagen, author of the fascinatingly and problematically hagiographical *South America Called Them: Explorations of the Great Naturalists La Condamine, Humboldt, Darwin, Spruce* (New York: Knopf, 1945).

2. Charles Kingsley, *At Last: A Christmas in the West Indies* (1871), vol. XIV of *The Works of Charles Kingsley*, 28 vols. (London: Macmillan, 1880–85), 387–88. Subsequent citations appear in the text following the abbreviation *AL*. Kingsley employs the first-person plural ("Up the Oroonoco we longed to go") because his daughter Rose accompanied him on this trip. The two left England together on December 2, 1869 on board the steamship *Shannon*.

3. In this Kingsley repeats earlier conceptions of Trinidad as the entrance to the mouth of the Orinoco and so the first step on the route to El Dorado – most notably those of Ralegh (detailed by Kingsley himself in "Sir Walter Raleigh and His Time" [1859], reprinted in *The Works of Charles Kingsley*,

xvi: 81–207) and of the Spanish conquistadors before him. See also V. S. Naipaul, *The Loss of El Dorado: A History* (New York: Knopf, 1970).

4. Christopher GoGwilt provides an extended treatment of the idea of "the West" around the turn of the nineteenth century in *The Invention of the West: Joseph Conrad and the Double-Mapping of Europe and Empire* (Stanford, CA: Stanford University Press, 1995).

5. For two recent and compelling treatments of the nineteenth-century British understanding of the tropics as a locus of illness, see Alan Bewell, *Romanticism and Colonial Disease* (Baltimore, MD: Johns Hopkins University Press, 2000) and Nancy Leys Stepan, *Picturing Tropical Nature* (Ithaca, NY: Cornell University Press, 2001).

6. Pierre Nora, "Between Memory and History: *Les Lieux de Mémoire*," *Representations* 26 (1989): 7–24. I discuss this term and my adoption of it in the Introduction.

7. Robert Bernard Martin, *The Dust of Combat: A Life of Charles Kingsley* (New York: Norton, 1960), 20–27.

8. Such indissociability of fact and fiction sometimes finds its way into biographers' accounts. Margaret Farrand Thorp, for example, writes: "Few travellers can have visited them [the West Indies] so well documented in advance. He knew stories of every bay and island, adventures which had befallen Elizabethan voyagers, and his own grandfather, adventures which he had invented for the personages of *Westward Ho!*" – as though his familiarity with the plot of one of his own novels constituted evidence of Kingsley's thorough grounding in the history and geography of the Americas. *Charles Kingsley, 1819–1875* (Princeton, NJ: Princeton University Press; London: Oxford University Press, 1937), 182.

9. This claim stands, I think, despite Kingsley's matter-of-fact assertion that "Crusoe's Island is almost certainly meant for Tobago" (*AL* 68).

10. On the New World – and specifically the New World tropics – as the site of a feared but desired libidinal release, see James Eli Adams, *Dandies and Desert Saints: Styles of Victorian Masculinity* (Ithaca, NY and London: Cornell University Press, 1995); Christopher Herbert, *Culture and Anomie: Ethnographic Imagination in the Nineteenth Century* (Chicago and London: University of Chicago Press, 1991); Stepan, *Picturing Tropical Nature*; and Robert J. C. Young, *Colonial Desire: Hybridity in Theory, Culture, and Race* (London and New York: Routledge, 1995).

11. On Britain and slavery in Brazil, see Leslie Bethell, *The Abolition of the Brazilian Slave Trade: Britain, Brazil, and the Slave Trade Question, 1807–1869* (Cambridge: Cambridge University Press, 1970); on the Eyre case, see Catherine Hall, *White, Male, and Middle-Class: Explorations in Feminism and History* (Cambridge: Polity, 1992), and Douglas Lorimer, *Colour, Class, and the Victorians: English Attitudes to the Negro in the Mid-Nineteenth Century* (Leicester: Leicester University Press, 1978). Kingsley himself subscribed to the Eyre Defence Committee, whose members, including Thomas Carlyle, Charles Dickens, John Ruskin, Alfred Tennyson, and John Tyndall,

believed Eyre's declaration of martial law and his execution of several hundred black Jamaicans justified. Darwin, Lyell, Huxley, John Stuart Mill, and others, by contrast, sought to have Eyre prosecuted for murder. J. M. I. Klaver, *The Apostle of the Flesh: A Critical Life of Charles Kingsley* (Leiden and Boston: Brill, 2006), 565–70.

12. Trollope also accomplishes the yoking together of the Caribbean with the Central and South American mainland by way of the map that stands as the frontispiece to his book, a map that stretches from the Windward Islands in the east to Guatemala in the west, from Cuba and the Yucatan peninsula in the north to New Granada (after 1903, Panama and Colombia) in the south. (Compare John R. Seeley, who writes of "the West Indian islands, among which I include some territories on the continent of Central and Southern America." *The Expansion of England* (1883; reprint, Chicago and London: University of Chicago Press, 1971), 14.) The difference from Kingsley's *At Last* and other texts under scrutiny in this chapter inheres in Trollope's studied indifference to the history of the lands he visits – encoded, among other places, in his declaration of the simultaneity of experiencing and writing about his journey: "I am beginning to write this book on board the brig – , trading between Kingston, in Jamaica, and Cien Fuegos, on the southern coast of Cuba." Anthony Trollope, *The West Indies and the Spanish Main* (second edition, 1860; reprint, London: Cass, 1968), 1. As Simon Gikandi writes of this opening: "[C]onceived as a space of writing, the moment of travel comes to represent a discursive and conceptual break with the past. And nothing illustrates this desire for a break from previous discourse better than Trollope's decision to make the scene of writing contiguous with the voyage itself." *Maps of Englishness: Writing Identity in the Culture of Colonialism* (New York: Columbia University Press, 1996), 92. Like Kingsley, however, Trollope can at least momentarily envision making the Americas his home: "When I settle out of England, and take to the colonies for good and all, British Guiana shall be the land of my adoption." Trollope, *The West Indies*, 169.

13. Ross Forman has argued that China and Hong Kong constitute other spaces resistant to Victorian imperial narratives as well as current scholarly accounts of Victorian imperialism; see "Projecting from Possession Point: Hong Kong, Hybridity, and the Shifting Grounds of Imperialism in James Dalziel's Turn-of-the-Century Fiction," *Criticism* 46.4 (Fall 2004): 533–74.

14. On belatedness and empire, see Ali Behdad, *Belated Travelers: Orientalism in the Age of Colonial Dissolution* (Durham, NC and London: Duke University Press, 1994), and Chris Bongie, *Exotic Memories: Literature, Colonialism, and the Fin de Siècle* (Stanford, CA: Stanford University Press, 1991).

15. Representative are Darwin's exclamations in the *Journal of Researches*:

We met, during our descent [of the part of the Paraná that runs through Argentina], very few vessels. One of the best gifts of nature seems here wilfully thrown away, in so grand a channel of communication being left unoccupied … How different would have been the aspect of this river, if English colonists had by good fortune first sailed up the Plata! What noble towns would now have occupied its shores! (Charles Darwin, *Journal of Researches into the Natural History and Geology of the Various Countries Visited by H.M.S. Beagle, under the Command of Captain FitzRoy, R.N, from 1832 to 1836* (1839), vols. ɪɪ and ɪɪɪ of *The Works of Charles Darwin*, 29 vols. (London: Pickering, 1986–89), ɪɪ: 128)

16. Charles Wentworth Dilke popularized the term in *Greater Britain: A Record of Travel in English-Speaking Countries During 1866–7*, 2 vols. (1868; reprint, London: Macmillan, 1869). Curiously, Dilke does not visit the West Indies. Apart from the few pages devoted to Mexico, Latin America figures in *Greater Britain* only in the text's conclusion, which among other improbabilities confidently predicts that "Chili, La Plata, and Peru must eventually become English." *Ibid.*, ɪ: 233–38, ɪɪ: 347.

17. Neil L. Whitehead, "South America/Amazonia: The Forest of Marvels," in *The Cambridge Companion to Travel Writing*, ed. Peter Hulme and Tim Youngs (Cambridge: Cambridge University Press, 2003), 137.

18. Compare Robert J. C. Young: "Perhaps the fixity of identity for which Englishness developed such a reputation arose because it was in fact continually being contested, and was rather designed to mask its uncertainty, its sense of being estranged from itself, sick with desire for the other." Young, *Colonial Desire*, 2. See also Ian Baucom, *Out of Place: Englishness, Empire, and the Locations of Identity* (Princeton, NJ: Princeton University Press, 1999).

19. Johannes Fabian defines "denial of coevalness" as *a persistent and systematic tendency to place the referent(s) of anthropology in a Time other than the present of the producer of anthropological discourse.*" *Time and the Other: How Anthropology Makes Its Object* (New York: Columbia University Press, 1983), 31. For a contestation of Fabian's claims, see Michael A. Elliott, "Other Times: Herman Melville, Lewis Henry Morgan, and Ethnographic Writing in the Antebellum United States," *Criticism* 49.4 (Fall 2007): 481–503.

20. Anne McClintock, *Imperial Leather: Race, Gender, and Sexuality in the Colonial Contest* (New York and London: Routledge, 1995), 30.

21. Joseph Conrad, *Heart of Darkness* (1899; reprint, Harmondsworth: Penguin, 1995), 62.

22. *Ibid.*, 54.

23. P. J. Cain and A. G. Hopkins, *British Imperialism: Innovation and Expansion 1688–1914* (London and New York: Longman, 1993), 312.

24. *Ibid.*, 276.

25. Leslie Bethell, referring to formal seizure of territory and usurpation of sovereignty, notes that "Latin America remained the only area of the globe largely free of imperialism in the nineteenth century." "Britain and Latin America in Historical Perspective," in *Britain and Latin America: A Changing Relationship*, ed. Victor Bulmer-Thomas (Cambridge and New York: Cambridge University Press, 1989), 13. The nature and extent of "informal imperialism" in the region continue to be subject to ongoing debate. Key documents include John Gallagher and Ronald Robinson, "The Imperialism of Free Trade," *Economic History Review* 6 (1953): 1–15; the

essays collected in D. C. M. Platt, ed., *Business Imperialism, 1840–1930: An Inquiry Based on British Experience in Latin America* (Oxford: Clarendon, 1977); Robert Freeman Smith, "Latin America, the United States and the European Powers, 1830–1930," in *The Cambridge History of Latin America*, ed. Leslie Bethell, 11 vols. (Cambridge: Cambridge University Press, 1984–94), IV: 83–119; the ninth chapter of Cain and Hopkins, *British Imperialism*, "Calling the New World into Existence: South America, 1815–1914"; Alan Knight, "Britain and Latin America," in *The Oxford History of the British Empire: The Nineteenth Century*, ed. Andrew Porter, 5 vols. (Oxford: Oxford University Press, 1999), III: 122–45; the essays in Raymond E. Dumett, ed., *Gentlemanly Capitalism and British Imperialism: The New Debate on Empire* (London and New York: Longman, 1999); Robert Aguirre, *Informal Empire: Mexico and Central America in Victorian Culture* (Minneapolis and London: University of Minnesota Press, 2005); and, in connection specifically with China but with implications for Latin America, Jürgen Osterhammel, "Semi-Colonialism and Informal Empire in Twentieth-Century China: Towards a Framework of Analysis," in *Imperialism and After: Continuities and Discontinuities*, ed. Wolfgang J. Mommsen and Osterhammel (London: German Historical Institute/Allen & Unwin, 1986), 290–314. Despite their many differences from one another, all sides of the debate at least confirm Catherine Hall's contention that "It is not possible to make sense of empire either theoretically or empirically through a binary lens." *Civilising Subjects: Metropole and Colony in the English Imagination 1830–1867* (Chicago and London: University of Chicago Press, 2002), 16. I discuss some of the implications of refusing binarism as a rubric for analyzing empire in "'The Sun and Moon Were Made to Give Them Light': Empire in the Victorian Novel," in *The Concise Companion to the Victorian Novel*, ed. Francis O'Gorman (London: Basil Blackwell, 2004), 4–24.

26. On Conrad and informal imperialism in South America, see my "Rumor, Shares, and Novelistic Form: Joseph Conrad's *Nostromo*," in *Victorian Investments: New Perspectives on Finance and Culture*, ed. Nancy Henry and Cannon Schmitt (Bloomington and Indianapolis: Indiana University Press, 2008), 182–201.

27. A partial list of those texts would include novels such as Kingsley's own *Westward Ho!*, Robert Louis Stevenson's *Treasure Island* (1883), and G. A. Henty's *Under Drake's Flag: A Tale of the Spanish Main* (1883); narratives of contemporary travel and exploration, such as Kingsley's *At Last* and J. A. Froude's *The English in the West Indies* (1888); and revisionist histories, such as Kingsley's "Sir Walter Raleigh and His Time" (1859) and James Rodway's *The West Indies and the Spanish Main* (1896).

28. Deborah Wormell, *Sir John Seeley and the Uses of History* (Cambridge: Cambridge University Press, 1988), 154–55; John Gross, "Editor's Introduction," in *The Expansion of England*, by Sir John R. Seeley (1883; reprint, Chicago and London: University of Chicago Press, 1971), xi–xii.

29. Seeley, *The Expansion of England*, 12. Subsequent citations appear in the text following the abbreviation *EE*. Anthony Pagden attributes Seeley's ability to imagine a Greater Britain to technological developments: "By the time Robert Seeley came to write *The Expansion of England* in 1883 … modern technologies of communication had shrunk the Atlantic Ocean, which for Burke had seemed an insuperable obstacle to federation. For Seeley, the Atlantic seemed 'scarcely broader than the sea between Greece and Sicily.' It was now possible to envisage a truly 'Greater Britain,' one that would constitute a global state." "The Empire's New Clothes: From Empire to Federation, Yesterday and Today," *Common Knowledge* 12.1 (Winter 2006): 44–45.

30. Catherine Hall cites Seeley's *Expansion of England* to demonstrate, in connection with the rethinking of colonialism necessitated by the postcolonial moment, the continued centrality but shifting signification of "race, nation and empire." "Introduction," in *Cultures of Empire: Colonizers in Britain and the Empire in the Nineteenth and Twentieth Centuries. A Reader*, ed. Hall (New York: Routledge, 2000), 2. Whereas for Seeley, Hall contends, these three categories or entities related to one another quite clearly (race demarcated those who belonged or did not belong to the nation; the nation, in the case of Great Britain, was constitutively imperial), such clarity is now complicated by a new history of empire, a re-theorization of racial and other differences as historically constructed, and especially a recognition that culture played (and continues to play) a crucial role in the maintenance of imperial rule and of racial difference. *Ibid.*, 10–18.

31. Wormell, *Sir John Seeley*, 155.

32. Gérard Genette, *Narrative Discourse: An Essay in Method*, trans. Jane E. Lewin (1972; English translation, Ithaca, NY: Cornell University Press, 1980), 40.

33. Here and throughout I follow Seeley in using "England" and "Britain" more or less interchangeably. In doing so, I do not wish to give the impression that the nomenclatural issues involved are anything but vexed and weighty. One effect of Seeley's usage is, of course, to marginalize Wales, Scotland, and Ireland by suggesting that the essence of Britain is English – and this is perhaps in accord with his concomitant attempt to centralize, as it were, New World colonial possessions. For the beginnings of a bibliography on this question, see Linda Colley, *Britons: Forging the Nation, 1707–1837* (New Haven, CT and London: Yale University Press, 1992); Robert Colls and Philip Dodd, eds., *Englishness: Politics and Culture, 1880–1920* (London: Croom Helm, 1986); and Tom Nairn, *The Break-up of Britain: Crisis and Neo-nationalism* (London: New Left Books, 1977).

34. Canning quoted in William W. Kaufmann, *British Policy and the Independence of Latin America, 1804–1828* (New Haven, CT: Yale University Press, 1951), 178.

35. Charles Kingsley, *Westward Ho!; or The Voyages and Adventures of Sir Amyas Leigh of Burrough, in the County of Devon, in the Reign of Her Most Glorious Majesty Queen Elizabeth*, vol. III of *The Works of Charles Kingsley*.

36. Most recent treatments of Kingsley focus on questions of masculinity and sexuality, often in connection with his position as an antagonist of the Oxford Movement. See among other works David Alderson, *Mansex Fine: Religion, Manliness and Imperialism in Nineteenth-Century British Culture* (Manchester: Manchester University Press, 1998); Charles Barker, "Erotic Martyrdom: Kingsley's Sexuality Beyond Sex," *Victorian Studies* 44.3 (Spring 2002): 465–88; Donald E. Hall, *Muscular Christianity: Embodying the Victorian Age* (Cambridge: Cambridge University Press, 1994); John Maynard, *Victorian Discourses on Sexuality and Religion* (Cambridge: Cambridge University Press, 1993); and Norman Vance, *The Sinews of the Spirit: The Ideal of Christian Manliness in Victorian Literature and Religious Thought* (Cambridge and New York: Cambridge University Press, 1985).

37. David Armitage, *The Ideological Origins of the British Empire* (Cambridge and New York: Cambridge University Press, 2000), 100.

38. Mary Louise Pratt, *Imperial Eyes: Travel Writing and Transculturation* (London and New York: Routledge, 1992), 27–28, 38–39.

39. See also the penultimate chapter, "A Provision Ground," which is entirely given over to a discussion of West Indian agriculture (*AL* 372–86).

40. Susan Chitty, *The Beast and the Monk: A Life of Charles Kingsley* (New York: Mason/Charter, 1975), 15.

41. Gikandi, *Maps of Englishness*, 98.

42. In this regard, see also Kingsley's contribution to the raft of popular guides to sea-side naturalizing that appeared toward the middle of the nineteenth century, *Glaucus; or, The Wonders of the Shore* (Boston: Ticknor and Fields, 1855), in which he writes: "[T]he dilettante (and it is for the dilettanti, like myself, that I principally write) must be content to tread in the tracks of greater men who have preceded him, and accept at second or third hand their foregone conclusions" (25). But this state of affairs is all to the good, he claims, for "the pleasure of finding new species is too great; it is morally dangerous; for it brings with it the temptation to look on the thing found as your own possession, all but your own creation; to pride yourself on it, as if God had not known it for ages since" (27–28). For a chapter-length treatment of *Glaucus*, see "Charles Kingsley and the Wonders of the Shore," the ninth chapter of Lynn Merrill's *The Romance of Victorian Natural History* (New York: Oxford University Press, 1989).

43. D. Graham Burnett, *Masters of All They Surveyed: Exploration, Geography, and a British El Dorado* (Chicago and London: University of Chicago Press, 2000), 39. Burnett's metalepsis and what I am calling analepsis may usefully be considered two versions of Romantic mnemonics, which James K. Chandler glosses as "the problem of the past pitted against the present for determinative priority." "About Loss: W. G. Sebald's Romantic Art of Memory," *SAQ* 102.1 (Winter 2003): 250. Chandler borrows here from Richard Terdiman, *Present Past: Modernity and the Memory Crisis* (Ithaca, NY: Cornell University Press, 1993), 84–85.

44. Burnett, *Masters of All They Surveyed*, 39.

45. William Wordsworth, *The Prelude* (1850; reprint, New York and London: W. W. Norton and Company, 1979), Book First, lines 269–74.

46. *Ibid.*, line 147.

47. *Ibid.*, Book Fourteenth, lines 303–04.

48. James K. Chandler, *Wordsworth's Second Nature: A Study of the Poetry and the Politics* (Chicago and London: University of Chicago Press, 1984), 187.

49. Seeley treats the metaphor of a nation's boyhood and maturity directly: "Greater Britain compared to old England may seem but the full-grown giant developed out of the sturdy boy; but there is this difference, that the grown man does not and cannot think of becoming a boy again, whereas England both can and does consider the expediency of emancipating her colonies and abandoning India." *EE* 132.

50. With the notable exception of Charles Darwin, this list features the principal European naturalists to have explored Latin America as of 1870. For detailed accounts of many of these figures, see Peter Raby, *Bright Paradise: Victorian Scientific Travellers* (London: Chatto and Windus, 1996); Anthony Smith, *Explorers of the Amazon* (New York: Viking, 1990); Von Hagen, *South America Called Them*; Whitehead, "South America/Amazonia: The Forest of Marvels." On natural history in the period, see David Elliston Allen, *The Naturalist in Britain: A Social History* (1976; reprint, Harmondsworth: Penguin, 1978); Nicholas Jardine, James A. Secord, and Emma Spary, eds., *Cultures of Natural History* (Cambridge: Cambridge University Press, 1996); Lynn Barber, *The Heyday of Natural History, 1820–1870* (London: Cape, 1980); and Merrill, *The Romance of Victorian Natural History*. J. M. I. Klaver writes of this aspect of *At Last*: "His [Kingsley's] own ancestors become part of England's heroic past in whose continuity Kingsley pictured himself as the explorer of natural history, or rather, in the person of a voyaging scientist like Darwin or Huxley, who, as he had repeatedly pointed out in his writings, were the modern equivalents of the English hero." *Apostle of the Flesh*, 608.

51. "For what the Orientalist does is to *confirm* the Orient in his readers' eyes; he neither tries nor wants to unsettle already firm convictions." Edward Said, *Orientalism* (1978; reprint, New York: Vintage, 1979), 65.

52. Palms were an obsession among the Victorians. Typical is the following, from an anonymous essay on "The Palms of Tropical America": "Thus when palm-trees figure so largely in the descriptions of tropical America given by travellers it is not without good reason. They impart an indefinable grace and charm to every landscape, and enter, in a thousand ways, into the daily occupations, thoughts, and feelings of the inhabitants." "The Palms of Tropical America," in *Illustrated Travels: A Record of Discovery, Geography, and Adventure*, ed. Henry Walter Bates, 2 vols. (London: Cassell, Petter, and Galpin, n.d.), II: 128. In a letter home written during his time in the Caribbean, Kingsley declared more emphatically: "so stands the palm tree, to be worshipped rather than loved." Qtd. in Martin, *Dust of Combat*, 271.

53. A reading of *At Last* together with Tennyson's "Locksley Hall," the scene of Amyas Leigh's temptation in *Westward Ho!*, and similar fantasies of escape to the tropics would suggest the place of the threat of miscegenation in keeping this longed-for idyll in the realm of the potential. Indeed, the supposedly dangerous attractions of dark women are alluded to throughout *At Last*, and often in connection with actual or impending disorder – as at a gathering in which "officers in uniform dance at desperate sailors' pace with delicate Creoles, some of them, colored as well as white, so beautiful in face and figure that one could almost pardon the jolly tars if they enacted a second Mutiny of the Bounty, and refused one and all to leave the island and the fair dames thereof" (*AL* 92). In his virtuoso reading of Kingsleyan tropical fantasy, James Eli Adams argues that in passages such as this one Kingsley "envision[s] the tropics as at once an extension of, and a site of resistance to, the demands of Victorian discipline." *Dandies and Desert Saints*, 123. Here the invocation of the fate of HMS *Bounty* suits perfectly: Fletcher Christian's mutiny links rejection of insupportable bodily discipline aboard William Bligh's ship with the lure of tropical sexuality. As for the *Bounty* mutineers and the speaker of "Locksley Hall," for Kingsley making the tropics home seems necessarily to entail the mixing of blood lines and the founding of a new, hybrid race.

54. Kingsley's more circuitous connections to Latin America include his friend Charles Mansfield, author of the posthumously published *Paraguay, Brazil, and the Plate* (1856), for which Kingsley wrote an introductory "Sketch of the Author's Life," and Kingsley's son Maurice, who spent time in Argentina (Chitty, *The Beast and the Monk*, 268; Martin, *Dust of Combat*, 275).

55. Compare Seeley, who writes of the settler colonies: "They have no past and an unbounded future." By contrast, "India is all past and, I may almost say, no future." *EE* 140, 141.

56. J. A. Froude, *The English in the West Indies; or, The Bow of Ulysses* (London: Longmans, Green, and Co., 1888), 72. To confirm that for Froude, too, the answer to the question posed to British imperialism by the West Indies and Spanish Main must be located in the past, in history and memory, we need only turn to J. J. Thomas's forthright rejoinder to *The English in the West Indies*, *Froudacity*, an early passage of which reads:

> Up to the date of the suggestion by [Froude] as above of the alleged facts and possibilities of West Indian life, we had believed (even granting the correctness of his gloomy account of the past and present positions of the two races) that to no well-thinking West Indian White, whose ancestors may have, innocently or culpably, participated in the gains as well as the guilt of slavery, would the remembrance of its palmy days be otherwise than one of mild regret. (*Froudacity: West Indian Fables by James Anthony Froude* (1888; second edition, London: T. Fisher Unwin, 1889), 9)

Many observations might be made about this passage. Here I will note only that Thomas's use of "palmy days" to refer to the period of (white) prosperity before the abolition of slavery makes available a disturbing gloss on Kingsley's palmy frontispiece. Thomas was a linguist; Kingsley refers to him in *At Last* when discussing Caribbean patois, mentioning a "curious book on it … by Mr. Thomas, a colored gentleman, who seems to be at once no mean philologer and no mean humorist" (*AL* 320). On Froude and Thomas, see also Gikandi, *Maps of Englishness*. On *The English in the West Indies* as an attempt to secure an imperial future by way of the Caribbean (and Homeric) past, see Tobias Döring, "The Sea is History: Historicizing the Homeric Sea in Victorian Passages," in *Fictions of the Sea: Critical Perspectives on the Ocean in British Literature and Culture*, ed. Bernhard Klein (Aldershot, Hampshire and Burlington, VT: Ashgate, 2002), 121–40.

57. Charles Kingsley to Charles Darwin, November 18, 1859. Frederick Burkhardt, Sydney Smith, *et al.*, eds., *The Correspondence of Charles Darwin* 16 vols. (Cambridge: Cambridge University Press, 1985–), VII: 380. Darwin received permission from Kingsley to include a modified version of this sentiment in the second edition of the *Origin*. *Ibid.*, 380n. The passage in question reads thus:

> I see no good reason why the views given in this volume should shock the religious feelings of any one. A celebrated author and divine has written to me that "he has gradually learnt to see that it is just as noble a conception of the Deity to believe that He created a few original forms capable of self-development into other and needful forms, as to believe that He required a fresh act of creation to supply the voids caused by the action of His laws." (Charles Darwin, *On the Origin of Species by Means of Natural Selection, or the Preservation of Favoured Races in the Struggle for Life* (1859; second edition, London: John Murray, 1860), 481)

58. Martin observes that Darwin's *Origin* "was accepted at once by Kingsley," and that Kingsley and T.H. Huxley maintained a lengthy correspondence "about the need for well-grounded studies in the new sciences for nineteenth-century Englishmen." *Dust of Combat*, 223, 274. Kingsley himself wrote to F.D. Maurice in 1863:

> I am very busy working out points of Natural Theology by the strange light of Huxley, Darwin, and Lyell … But I am not going to rush into print this seven years, for this reason: the state of the scientific world is most curious; Darwin is conquering everywhere, and rushing in like a flood by the mere force of truth and fact. (Quoted in Guy Kendall, *Charles Kingsley and His Ideas* (London: Hutchinson and Co., n.d.), 135)

59. On *The Water-Babies* in relation to *On the Origin of Species*, see Gillian Beer, *Darwin's Plots: Evolutionary Narrative in Darwin, George Eliot and Nineteenth-Century Fiction* (1983; revised edition, Cambridge and New York: Cambridge University Press, 2000), 114–29.

60. Charles Kingsley, "The Natural Theology of the Future" (1871), reprinted in *Scientific Lectures and Essays* (1880; revised edition, London and New York: Macmillan and Co., 1890), 331.

61. The second quotation appears at the end of a passage in which Kingsley cites the *Origin*:

> We were taught, some of us at least, by Holy Scripture, to believe that the whole history of the universe was made up of special providences: if, then, that should be true which Mr. Darwin says – "It may be metaphorically said that natural selection

is daily and hourly scrutinizing, throughout the world, every variation, even the slightest; rejecting that which is bad, preserving and adding up all that is good; silently and insensibly working, whenever and wherever opportunity offers, at the improvement of each organic being in relation to its organic and inorganic conditions of life," – if this, I say, were proved to be true, ought God's care, God's providence, to seem less or more magnificent in our eyes? (*AL* 246)

62. For a succinct account of the debate, see Robert J. Richards, *Darwin and the Emergence of Evolutionary Theories of Mind and Behavior* (Chicago and London: University of Chicago Press, 1987), 3–4.
63. Charles Kingsley to Charles Darwin, January 31, 1862. Burkhardt *et al.*, *Correspondence* x: 63.
64. *Ibid.*, 63–64.
65. Darwin, *Origin of Species*, 470.
66. Charles Darwin to Charles Kingsley, February 6, 1862. Burkhardt *et al.*, *Correspondence* x: 71.
67. Darwin does, however, opine that "[i]t is very true what you say about the higher races of men, when high enough, replacing & clearing off the lower races." He also speaks directly to the question of the dove, declaring with the offhand certainty of the professional:

With respect to the pigeons, your remarks show me clearly … that the birds shot were the Stock Dove or C. Oenas, long confounded with the Cushat & Rock-pigeon. It is in some respects intermediate in appearance & habits; as it breeds in *holes* in trees & in rabbit-warrens. It is so far intermediate that it quite justifies what you say on all the forms being descendants of one. (*Ibid.*, 72, 71)

68. Thomas Henry Huxley, *Evidence as to Man's Place in Nature* (London and Edinburgh: Williams and Norgate, 1863), 1.
69. See also the following repellant aphorism: "If any one says of the negro, as of the Russian, 'He is but a savage polished over: you have only to scratch him, and the barbarian shows underneath'; the only answer to be made is, Then do not scratch him" (*AL* 88–89).
70. Charles Kingsley to Charles Darwin, January 31, 1862. Frederick Burkhardt *et al.*, *Correspondence* x: 63.
71. Kingsley, *Glaucus*, 14–15.

4 W. H. HUDSON'S MEMORY OF LOSS

1. Edward Garnett, "The Genius of W. H. Hudson," in *A Hind in Richmond Park*, by W. H. Hudson (New York: Dutton, 1923), xiii. A somewhat abbreviated version of Garnett's essay appears in the last volume of the *Collected Works* under the title "A Note on Hudson's Spirit." See W. H. Hudson, *A Traveller in Little Things*, volume xxiv of *The Collected Works of W. H. Hudson*, 24 vols. (New York: Dutton, 1922–23), vii–xiii.
2. W. H. Hudson to Alice Rothenstein, August 15, 1906. W. H. Hudson, *The Unpublished Letters of W. H. Hudson, The First Literary Environmentalist 1841–1922*, ed. Dennis Shrubsall, 2 vols. (Lewiston, NY, Queenston, ON, and

Lampeter, Ceredigion: The Edwin Mellen Press, 2006), II: 402. For another obituary depiction of Hudson as primitive, see the antepenultimate paragraph of Morley Roberts's biography:

> I stayed with him a long while. Presently they would bury him. Such a man! Such a beautiful big savage and genius – to be buried as we bury! I wished to take him out upon the open pampa, with a long wide view beyond the sight of man even on horseback, with the great clear sky above. So I would have digged a grave and put him there to rest in his blanket just as he had fallen asleep, without disturbing his attitude of quiet peace. (Morley Roberts, *W. H. Hudson: A Portrait* (New York: E. P. Dutton and Company, 1924), 305)

3. W. H. Hudson, *Far Away and Long Ago: A Childhood in Argentina* (1918; reprint, London: Eland Books, 1982), 225. Subsequent citations appear in the text following the abbreviation *FA*.

4. Garnett, "The Genius of W. H. Hudson," xvii, xxii. The references are to Henry Walter Bates (1825–92), author of *The Naturalist on the River Amazons* (1863); Alfred Russel Wallace, the subject of the second chapter of this book; Thomas Belt (1832–78), author of *The Naturalist in Nicaragua* (1874); Richard Jefferies (1848–87), author of among other books *Round About a Great Estate* (1880), *The Story of My Heart* (1883), and *After London* (1885); and John Burroughs (1837–1921), author of many books of natural history including *Wake-Robin* (1871).

5. Alfred Russel Wallace, "A Man of the Time: Dr. Alfred Russel Wallace and His Coming Autobiography," an interview by "J. M.," *The Book Monthly* (May 1905): 548–49. This laudatory judgement repeats that of his earlier review of *The Naturalist in La Plata*:

> Never has the present writer derived as much pleasure and instruction from a book on the habits and instincts of animals. He feels sure that it will long continue to be a storehouse of facts and observations of the greatest value to the philosophical naturalist, while to the general reader it will rank as the most interesting and delightful of modern books on natural history. (Alfred Russel Wallace, "A Remarkable Book on the Habits of Animals," *Nature* 45 [April 14, 1892]: 556)

6. Ford Madox Ford, *Mightier than the Sword* (London: Allen & Unwin, 1938), 67. Garnett's and Ford's estimates are representative rather than anomalous, as David Miller establishes in *W. H. Hudson and the Elusive Paradise* (New York: St. Martin's, 1990), 1–6.

7. Hudson's biographer Ruth Tomalin writes: "He would later declare that his life ended when he left the pampas." *W. H. Hudson: A Biography* (London: Faber and Faber, 1982), 136. See also Nicholas Shakespeare: "Hudson used to say that his life ended when he left South America." "Preface" to W. H. Hudson, *Far Away and Long Ago*, xv.

8. As Simon Collier puts it, "The tendrils of Hudson's initial thirty-three years, those of his first, Argentine, world, clung to him stubbornly through his last forty-eight, the years of his second, English, world." "The Four Worlds of

W.H. Hudson," in *The Borges Tradition*, ed. Norman Thomas di Giovanni (London: Constable, in association with the Anglo-Argentine Society, 1995), 73.

9. Hudson provides this chronology in *The Naturalist in La Plata*. In a campaign known as the Conquest of the Desert, General Julio A. Roca and his troops killed more than a thousand indigenous inhabitants of the pampas in 1879. In 1880, on the strength of his reputation after this slaughter, Roca was elected President of Argentina. Silently eliding the violence that made them possible, an economic historian itemizes some of the results:

> It was during the late 1870s that Argentina became a net exporter of grains, a trade which began on a small scale but which quickly gained momentum. Between 1872 and 1895, the amount of pampa acreage under cultivation in all crops, especially grains, grew fifteen times, and in the next decade, the amount of acreage planted in wheat and maize alone more than doubled. (William Glade, "Latin America and the International Economy, 1870–1914," in *The Cambridge History of Latin America*, ed. Leslie Bethell, 11 vols. (Cambridge: Cambridge University Press, 1984–94), IV: 11)

10. Ian Duncan, "Introduction," *Green Mansions: A Romance of the Tropical Forest*, by W.H. Hudson (1904; reprint, Oxford and New York: Oxford University Press, 1998), viii.

11. The RSPB's website chronicles some of Hudson's contributions to the organization during its first three decades: www.rspb.org.uk/about/history/milestones.asp.

12. Andreas Huyssen, *Present Pasts: Urban Palimpsests and the Politics of Memory* (Stanford, CA: Stanford University Press, 2003), 4.

13. The remaining titles by Hudson listed under the heading "Reminiscences of a Naturalist" are *Far Away and Long Ago*, *A Traveller in Little Things*, *A Shepherd's Life*, and *The Book of a Naturalist*. W.H. Hudson, *A Hind in Richmond Park* (New York: Dutton, 1923), front matter.

14. Huyssen, *Present Pasts*, 27.

15. *Ibid.*, 18, 3. In *Present Past: Modernity and the Memory Crisis* (Ithaca, NY: Cornell University Press, 1993), Richard Terdiman establishes just how far-reaching, geographically and conceptually, was the strain of "memory fever" that plagued nineteenth-century Europe. Frances Ferguson, discussing Wordsworth in particular, details the fateful "uses that romanticism develops for memory and the consequences that such uses of memory have for both history and the notion of individual identity." Frances Ferguson, "Romantic Memory," *Studies in Romanticism* 35.4 (1996): 510.

16. Cathy Caruth, "Trauma and Experience: Introduction," in *Trauma: Explorations in Memory*, ed. Caruth (Baltimore, MD and London: Johns Hopkins University Press, 1995), 4–5. See also Cathy Caruth, *Unclaimed Experience: Trauma, Narrative and History* (Baltimore, MD and London: Johns Hopkins University Press, 1996).

17. Nicholas Dames specifies the nature of the forgetting inherent in such a form of recollection when he writes that "[a] nostalgic looking-backward is … necessarily a looking-forward – a dilution and disconnection of the past in the service of an encroaching future." *Amnesiac Selves: Nostalgia,*

Forgetting, and British Fiction, 1810–1870 (Oxford and New York: Oxford University Press, 2001), 236. Under a regime of nostalgia, recollection sanitizes the past by selectively preserving and arranging only those elements of it that allow the construction of a coherent – and comforting – narrative about the present and future. Such was not Hudson's practice. The past was unfathomably magical, he insists, but also and always shot through with loss and impossible to speak of without reference to that loss. For a wonderfully dense historicization and exploration of nostalgia and its maritime inflections, see Jonathan Lamb, *Preserving the Self in the South Seas, 1680–1840* (Chicago and London: University of Chicago Press, 2001). See also Michael Mason, "The Cultivation of the Senses for Creative Nostalgia in the Essays of W. H. Hudson," *ARIEL: A Review of International English Literature* 20 (1989): 23–37, and Ann C. Colley, *Nostalgia and Recollection in Victorian Culture* (Basingstoke and London: Macmillan; New York: St. Martin's, 1998), the first chapter of which treats the place of nostalgia in Darwin's *Journal of Researches*.

18. W. H. Hudson, *A Hind in Richmond Park*, volume XXIII of *The Collected Works of W. H. Hudson*, xii. Subsequent citations appear in the text following the abbreviation *HRP*.

19. Charles Darwin, *The Descent of Man, and Selection in Relation to Sex*, 2 vols. (1871; reprint, Princeton, NJ: Princeton University Press, 1981), II: 404.

20. Gillian Beer, *Darwin's Plots: Evolutionary Narrative in Darwin, George Eliot and Nineteenth-Century Fiction* (1983; revised edition, Cambridge and New York: Cambridge University Press, 2000), 20.

21. This is also true of sexual selection, but the role of will or desire in sexual selection complicates its implications in this context: if, on one hand, both natural and sexual selection premise an absolutely contingent and unique biological world, on the other the place of the desires of the individual, revealed as irrelevant by natural selection, return with sexual selection in such a way as to make repeatability, return, thinkable again.

22. Although Hudson never completely accepted evolutionary theory, it is a mistake to overemphasize his disagreement with it. He himself provides the most measured and accurate assessment: "Insensibly and inevitably I had become an evolutionist, albeit never wholly satisfied with natural selection as the only and sufficient explanation of the change in forms of life." *FA* 330. Robert M. Young shows that such a response to Darwin was common; see *Darwin's Metaphor: Nature's Place in Victorian Culture* (Cambridge: Cambridge University Press, 1985). Even so eminent a Darwinian as Herbert Spencer, for instance, argued that Darwin "leaves unconsidered a mass of morphological phenomena that are explicable as results of functionally-acquired modifications, transmitted and increased, and which are not explicable as results of natural selection." *The Principles of Biology*, 2 vols. (London: Williams and Norgate, 1864), 1: 449. For an extended treatment of Hudson's relation to Darwinism as it plays itself out in one of Hudson's novels, see John Glendening, "Darwinian Entanglement in Hudson's *Green*

Mansions," *ELT* 43 (2000): 259–79. For views that stress Hudson's resistance to evolutionary theory, see Miller, *W. H. Hudson and the Elusive Paradise* throughout but especially 79–81, and Richard E. Haymaker, *From Pampas to Hedgerows and Downs: A Study of W. H. Hudson* (New York: Bookman, 1954), especially 197–98.

23. Marjorie Grene and David Depew, *The Philosophy of Biology: An Episodic History* (Cambridge and New York: Cambridge University Press, 2004), 214. Hudson registers the same insight more lyrically:

> And every creature I watched, from the great soaring bird circling in the sky at a vast altitude to the little life at my feet … was a type, representing a group marked by family likeness not only in figure and colouring and language, but in mind as well, in habits and the most trivial traits and tricks of gesture and so on … What explanation was possible but that of community of descent? (*FA* 329)

For further discussion of the implications of this familialism, see my essay "Victorian Beetlemania," in *Victorian Animal Dreams: Representations of Animals in Victorian Literature and Culture,* ed. Deborah Denenholz Morse and Martin A. Danahay (Aldershot, Hampshire and Burlington, VT: Ashgate, 2007), 35–51.

24. Ian Duncan writes about the dénouement of *Green Mansions*: "Internalization, the compensatory mechanism of romantic loss, yields a 'Rima of the mind' who eventually sanctions the bleak self-reliance of a man 'self-forgiven and self absolved.'" "Introduction," xxi. An enormous body of work exists that examines the history of European primitivism in relation to the question of "going native." An early account interesting especially for its taxonomy of primitivisms is Arthur O. Lovejoy's "Prolegomena to the History of Primitivism," in *Primitivism and Related Ideas in Antiquity,* by Lovejoy and George Boas (Baltimore, MD: Johns Hopkins University Press, 1935), 1–22. More recent treatments include, among many others, Chris Bongie, *Exotic Memories: Literature, Colonialism, and the Fin de Siècle* (Stanford, CA: Stanford University Press, 1991); Marianna Torgovnick, *Gone Primitive: Savage Intellects, Modern Lives* (Chicago and London: University of Chicago Press, 1990) and *Primitive Passions: Men, Women, and the Quest for Ecstasy* (Chicago and London: University of Chicago Press, 1996). Victor Li mounts a critique of Torgovnick and some of her contemporaries as lapsing into neo-primitivism in *The Neo-Primitivist Turn: Critical Reflections on Alterity, Culture, and Modernity* (Toronto: University of Toronto Press, 2006); see also Li's exchange with Torgovnick and Adam Kuper in *Criticism* 49.4 (Fall 2007): 545–63. On specifically Darwinian conceptions of the extinction of "primitives," see the eighth chapter of Patrick Brantlinger's *Dark Vanishings: Discourse on the Extinction of Primitive Races, 1800–1930* (Ithaca, NY and London: Cornell University Press, 2003), 164–88.

25. W. H. Hudson, *The Naturalist in La Plata* (1892), volume VIII of *The Collected Works of W. H. Hudson,* 2. Subsequent citations appear in the text following the abbreviation *NLP.*

26. W. H. Hudson, letter to the Secretary of the Zoological Society, London, *Proceedings of the Zoological Society*, Part 1 (1870): 158–60. Reprinted, with Darwin's reply, in Tomalin, *W. H. Hudson*, 237–42. The claim in question reads: "[O]n the plains of La Plata, where not a tree grows, there is a wood-pecker, which in every essential part of its organisation, even in its colouring, in the harsh tone of its voice, and undulatory flight, told me plainly of its close blood-relationship to our common species; yet it is a woodpecker which never climbs a tree!" Charles Darwin, *On the Origin of Species by Means of Natural Selection, or the Preservation of Favoured Races in the Struggle for Life* (1859; reprint, Cambridge, MA and London: Harvard University Press, 1964), 184.

27. William Shakespeare, *The Comedy of Errors*, in *The Riverside Shakespeare*, ed. G. Blakemore Evans (second edition; Boston: Houghton Mifflin, 1997), v.i.302–04.

28. Darwin, *Origin of Species*, 62.

29. See Gilbert White, *The Natural History and Antiquities of Selborne* (London: Printed by T. Bensley, for B. White and Son, 1789). Hudson acquired a copy of the book at the age of 15. Tomalin notes, "One cannot doubt that White's letters were the model for his own initial work, the foundation of his career as a writer: his letters to the Zoological Society of London." *W. H. Hudson*, 63. See also Haymaker, *From Pampas to Hedgerows and Downs*, 59–62.

30. My allusion is to the title of the English translation of the book of W. G. Sebald's essays originally published as *Luftkrieg und Literatur* [Air War and Literature] (1999): *On the Natural History of Destruction*, trans. Anthea Bell (New York: Modern Library, 2004).

31. Ezra Pound, "Hudson: Poet Strayed into Science," in *Selected Prose, 1909–1965* (London: Faber and Faber, 1978), 399. Compare one of Darwin's descriptions of indigenes of the Río de la Plata region, imagined in response to a story he was told about a "cacique" or chief who fled from an attack by Argentine troops: "The old Indian father and his son escaped, and were free. What a fine picture one can form in one's mind, the naked bronze-like figure of the old man with his little boy, riding like a Mazeppa on the white horse, thus leaving far behind him the host of this pursuers!" Charles Darwin, *Journal of Researches into the Natural History and Geology of the Various Countries Visited by H.M.S. Beagle, under the Command of Captain FitzRoy, R.N, from 1832 to 1836* (1839), vols. II and III, *The Works of Charles Darwin*, 29 vols. (London: Pickering, 1986–89), II: 95.

32. Duncan notes of this passage that "Hudson invests the Indians with an epiphany of the vanishing wilderness and his own loss; their fate can represent the general extinction he mourns." "Introduction" xxii.

33. W. H. Hudson, *NLP* (London: Chapman and Hall, 1892), frontispiece.

34. Compare a similar remark by Hudson in an October 17, 1893 letter to the *Times* titled "Feathered Women." Speculating on "our remote descendants," Hudson writes: "They will, I fancy, think less kindly of their cultured, Ruskin-reading nineteenth-century ancestors, than of those very much

more distant progenitors who had some shocking customs, but spoilt nothing." Qtd. in Tomalin, *W. H. Hudson*, 246.

35. Hudson's mention of "that little hairy maiden exhibited not long ago in London" may be a reference to Julia Pastrana, a native of Mexico exhibited in London in the late 1850s as a "Bear Woman" or "Nondescript." See Janet Browne and Sharon Messenger, "Victorian Spectacle: Julia Pastrana, the Bearded and Hairy Female," *Endeavour* 27.4 (December 2003): 155–59 and Rosemarie Garland Thomson, "Narratives of Deviance and Delight: Staring at Julia Pastrana, the 'Extraordinary Lady,'" in *Beyond the Binary: Reconstructing Cultural Identity in a Multicultural Context*, ed. Timothy B. Powell (New Brunswick, NJ: Rutgers University Press, 1999), 81–104.

36. Dana Seitler's *Atavistic Tendencies: The Culture of Science in American Modernity* (Minneapolis and London: University of Minnesota Press, 2008) provides the definitive account to date of atavism's complex political and aesthetic history.

37. Darwin writes:

> For my part, following out Lyell's metaphor, I look at the natural geological record, as a history of the world imperfectly kept, and written in a changing dialect; of this history we possess the last volume alone, relating only to two or three countries. Of this volume, only here and there a short chapter has been preserved; and of each page, only here and there a few lines. (*Origin of Species* 310–11)

38. Tim Fulford, Debbie Lee, and Peter J. Kitson, *Literature, Science and Exploration in the Romantic Era: Bodies of Knowledge* (Cambridge: Cambridge University Press, 2004), 14. The sixth chapter of their book, "Exploration, Headhunting and Race Theory: The Skull Beneath the Skin," provides a detailed examination of the scientific trade in humans and human remains.

39. I specify Hudson's natural history writing because there are notable moments in his first novel in which the protagonist, Richard Lamb, becomes savage as a consequence of his contact with savages. Lamb notes about his feelings on killing a man who persecuted him:

> [I]f a murderous brute with truculent eyes and gnashing teeth attempts to disembowel me with a butcher's knife, the instinct of self-preservation comes out in all its old original ferocity, inspiring the heart with such implacable fury that after spilling his blood I could spurn his loathsome carcass with my foot. I do not wonder at myself for speaking those savage words. [He had said to the dead man's companions: "I make you a present of his carcass."] That he was past recall seems certain, yet not a shade of regret did I feel at his death. Joy at the terrible retribution I had been able to inflict on the murderous wretch was the only emotion I experienced when galloping away into the darkness – such joy that I could have sung and shouted aloud had it not seemed imprudent to indulge in such expression of feeling. (W. H. Hudson, *The Purple Land: Being the Narrative of one Richard Lamb's Adventures in the Banda Oriental, in South America, as told by Himself* (second edition, 1904; reprint, Madison and London: University of Wisconsin Press, 2002), 176)

40. Hudson, *Green Mansions*, 177. At this point in the novel Abel, the narrator, identifies savages with natural selection itself; after exclaiming at the horror of Rima's murder he writes: "But I knew it all before – this law of nature and of necessity, against which all revolt is idle: often had the remembrance of it filled me with ineffable melancholy; only now it seemed cruel beyond all cruelty." *Ibid.*

41. On my adoption and adaptation of Pierre Nora's term *lieu de mémoire*, see the Introduction to the present volume.

42. W. H. Hudson, *Idle Days in Patagonia* (1893), volume xvi of *Collected Works of W. H. Hudson*, 103; emphasis in the original here and throughout unless otherwise indicated. Subsequent citations appear in the text following the abbreviation *IDP*.

43. This is, in essence, my argument in chapter 3, "Charles Kingsley's recollected empire."

44. Here and elsewhere, Hudson appears to be in continuous but unmarked dialogue with a towering Argentine predecessor, Domingo Faustino Sarmiento, who, in the book that would come to be known as *Civilization and Barbarism*, opined:

     Before 1810, two distinct, rival, and incompatible forms of society, two differing kinds of civilization existed in the Argentine Republic: one being Spanish, European, and cultivated, the other barbarous, American, and almost wholly of native growth. The revolution which occurred in the cities acted only as the cause, the impulse, which set these two distinct forms of national existence face to face, and gave occasion for a contest between them, to be ended, after lasting many years, by the absorption of one into the other. (*Life in the Argentine Republic in the Days of the Tyrants; or, Civilization and Barbarism*, trans. Mary Mann (1845; English translation from the third spanish edition, New York: Hurd and Houghton; Cambridge: Riverside Press, 1868), 54)

45. Darwin, *Descent of Man*, ii: 404.

46. Marcel Proust, *Swann's Way* (1913), trans. C. K. Scott Moncrieff (Harmondsworth: Penguin, 1957), 55. Such a memory is "involuntary" in the sense that it cannot be willed; nonetheless, as Proust's narrator carefully notes, effort is required if the memory is "ultimately [to] reach the clear surface of … consciousness" (57).

47. "Recapture" translates Proust's "évoquer": "Il en est ainsi de notre passé. C'est peine perdue que nous cherchions à l'évoquer, tous les efforts de notre intelligence sont inutiles." Marcel Proust, *A La recherche du temps perdu*, 3 vols. (Paris: Editions Gallimard, 1954), i: 44.

48. Proust, *Swann's Way*, 56.

49. Pierre Nora, "Between Memory and History: *Les Lieux de Mémoire*," *Representations* 26 (1989): 17.

50. This difference from Proust signals Hudson's proximity to Bergson, who writes:

     Whenever we are trying to recover a recollection … we become conscious of an act *sui generis* by which we detach ourselves from the present in order to replace ourselves, first, in the past in general, then, in a certain region of the past – a work of adjustment, something like the focusing of a camera. But our recollection still remains virtual; we simply prepare ourselves by adopting the appropriate attitude. (Henri Bergson, *Matter and Memory* (1896; fifth edition, 1908), trans. N M. Paul and W. S. Palmer (reprint; New York: Zone, 1991), 133–34)

Gilles Deleuze's comments on this passage further illuminate the Bergsonian nature of Hudson's return to the past: "if this act is *'sui generis,'* this is because it has made a genuine *leap*. We place ourselves *at once* in the past; we leap into the past as into a proper element." Gilles Deleuze, *Bergsonism*, trans. Hugh Tomlinson and Barbara Habberjam (1966; New York: Zone, 1991), 56.

51. Alfred Russel Wallace, "Reveries of a Naturalist," *Nature* (March 23, 1893): 483.

52. Compare a claim in *A Hind in Richmond Park*: "The vestiges of [the] past are numerous enough, and when collected and classified they may form a new subject or science with a specially invented new name, signifying an embryology of the mind" (*HRP* 52).

53. Diane Ackerman, *A Natural History of the Senses* (1990; reprint, New York: Vintage, 1995); Michael Taussig, *Mimesis and Alterity: A Particular History of the Senses* (New York and London: Routledge, 1993). See also several works by Alain Corbin, especially *Time, Desire and Horror: Towards a History of the Senses*, trans. Jean Birrell (Cambridge: Polity, 1995). Aspirations for such an investigation date back at least to Bacon's *Novum Organum* (1620); see Paul B. Wood, "The Science of Man," in *Cultures of Natural History*, ed. by N. Jardine, J.A. Secord, and E.C. Spary (Cambridge: Cambridge University Press, 1996), 197–98. For a discussion of recent efforts in this area that, like Hudson's, are indebted to an evolutionary perspective, see Kate Flint, "Sensuous Knowledge," in *Unmapped Countries: Biological Visions in Nineteenth Century Literature and Culture*, ed. Anne-Julia Zwierlein (London: Anthem, 2005), 207–15.

54. In note 22, above, I discuss Hudson's acceptance of evolution despite his rejection of natural selection as its mechanism.

55. Harold Goddard notes that the "titles of a number of [Hudson's books] are, consciously or unconsciously, symbolic: *A Hind in Richmond Park*, for instance. Hudson was that hind." *W. H. Hudson: Bird-Man* (New York: Dutton, 1928), 31. See also a later passage in *A Hind*, in which Hudson writes: "Little by little the knowledge comes that, notwithstanding the enormous difference between man and animals, mentally it is one of degree only, that all that is in our minds is also in theirs" (*HRP* 262).

56. The categories are of course derived from those of René Wellek and Austin Warren, "intrinsic" and "extrinsic" literary history; see *Theory of Literature* (New York: Harcourt, Brace, 1949).

57. Much of that incoherence derives from the fact that *A Hind in Richmond Park* was left unfinished at the time of Hudson's death. In a "Prefatory Note" to the edition of the book published in the *Collected Works*, Morley

Roberts itemizes the difficulties he faced in preparing the manuscript for publication (*HRP* vii–viii).

58. Compare *NLP*, chapter 19, "Music and Dancing in Nature," another instance of Hudson's attempt to refute Darwin's theory that the origins of music are to be found in sexual selection. Instead, Hudson attributes its development to something like a spontaneous overflow of animal vitality:

> We see that the inferior animals, when the conditions of life are favourable, are subject to periodical fits of gladness, affecting them powerfully and standing out in vivid contrast to their ordinary temper. And we know what this feeling is – this periodic intense elation which even civilised man occasionally experiences when in perfect health, more especially when young. There are moments when he is mad with joy, when he cannot keep still, when his impulse is to sing and shout aloud and laugh at nothing, to run and leap and exert himself in some extravagant way. (*NLP* 275)

59. This quotation stands as the epigraph to the *Collected Works* edition of *A Hind in Richmond Park*.

60. Stephen Jay Gould, *Wonderful Life: The Burgess Shale and the Nature of History* (New York and London: Norton, 1989), 24–25. Subsequent citations appear in the text following the abbreviation *WL*. For grasping just how strange the Burgess creatures were, nothing replaces the illustrations Gould commissioned for the book or the short animated film on view at the Royal Ontario Museum, Toronto. But consider part of Gould's prose description of the aptly named *Hallucigenia*:

> In broad outline, *Hallucigenia* has a bulbous "head" on one end ... We cannot even be certain that this structure represents the front of the animal; it is a "head" by convention only. This "head" ... attaches to a long, narrow, basically cylindrical trunk. Seven pairs of sharply pointed spines – not jointed, arthropod-like appendages, but single discrete structures – connect to the sides of the trunk ... Along the dorsal mid-line of the body, directly opposite the spines, seven tentacles with two-pronged tips extend upward ... A cluster of six much shorter dorsal tentacles (perhaps arranged as three pairs) lies just behind the main row of seven. The posterior end of the trunk then narrows into a tube and bends upward and forward. (*Ibid.*, 155)

61. As Gould notes, the different meanings of "diversity" may cause confusion; for the sake of clarity he proposes that "diversity" refer only to the number of species, "disparity" to the number of different anatomical plans. "Using this terminology, we may acknowledge a central and surprising fact of life's history – marked decrease in disparity followed by an outstanding increase in diversity within the few surviving designs." *Ibid.*, 49.

62. Walter Benjamin, "On Some Motifs in Baudelaire," in *Illuminations*, trans. Harry Zohn (New York: Schocken, 1968), 158.

63. Of course, for Hudson, Darwin, and many others the personal past and the species past were interimplicated insofar as the former could be understood to constitute a recapitulation of the latter. Hudson, for example, treating the evolution of art in *A Hind in Richmond Park*, claims: "[W]e can see all

the early stages [of artistic production] in our own young barbarians playing in a mud-puddle, progressing from printing a foot with all its little toes
complete to the moulding of 'mud-pies,' and so on till the period of drawing
human figures on a slate" (*HRP* 309). On one of the crucial twentieth-
century elaborations of this interimplication, that of Sigmund Freud, see
Dana Seitler, "Freud's Menagerie," *Genre* 38 (Spring/Summer 2005): 45–70;
see also Stephen Jay Gould, *Ontogeny and Phylogeny* (Cambridge, MA and
London: Harvard University Press, 1977), 155–64; Frank Sulloway, *Freud:
Biologist of the Mind* (New York: Basic Books, 1983), 258–64.

64. See Peter Mason, *Infelicities: Representations of the Exotic* (Baltimore, MD
and London: Johns Hopkins University Press, 1998) and *The Lives of Images*
(London: Reaktion, 2001); Kobena Mercer, *Welcome to the Jungle: New
Positions in Black Cultural Studies* (New York and London: Routledge,
1994); Anne McClintock, *Imperial Leather: Race, Gender, and Sexuality in
the Colonial Contest* (New York and London: Routledge, 1995); Edward Said,
Culture and Imperialism (New York: Knopf, 1993).

65. Jorge Luis Borges, "About *The Purple Land*," in *Other Inquisitions 1937–1952*,
trans. Ruth L. C. Simms (1952; English translation, Austin: University of
Texas Press, 1964), 144. Eva-Lynn Alicia Jagoe treats Borges's appropriations
of Hudson in *The End of the World as They Knew It: Writing Experiences of
the Argentine South* (Lewisburg, PA: Bucknell University Press, 2008).

CODA: SOME REFLECTIONS

1. This is typical of contemporary caricatures. As Jonathan Smith writes:
"Even when good natured, these cartoons relied for their humor on what
was clearly a widespread sense of Darwinism's fundamental absurdities –
that species could change abruptly, that such an obviously superior species as
humans could have developed from animals, that the great Darwin himself
was the descendant of apes." *Charles Darwin and Victorian Visual Culture*
(Cambridge and New York: Cambridge University Press, 2006), 235. See
also Gowan Dawson's illuminating discussion of caricatures of Darwin in
the wake of public disquiet over the seemingly obsessive treatment of sexual matters in *The Descent of Man, and Selection in Relation to Sex* (1871).
Darwin, Literature and Victorian Respectability (Cambridge and New York:
Cambridge University Press, 2007), 55–74.

2. Janet Browne, "Darwin in Caricature: A Study in the Popularisation and
Dissemination of Evolution," *Proceedings of the American Philosophical
Society* 145.4 (December 2001): 501. Browne claims that caricatures such as
"Prof. Darwin" contributed significantly to the popular understanding of
evolutionary theory in the nineteenth century.

3. Charles Darwin, *The Descent of Man, and Selection in Relation to Sex*, 2 vols.
(1871; reprint, Princeton, NJ: Princeton University Press, 1981), II: 389.

4. Pierre Nora, "Between Memory and History: *Les Lieux de Mémoire*,"
Representations 26 (1989): 17–18.

5. Frances Ferguson, "Romantic Memory," *Studies in Romanticism* 35.4 (1996): 509.

6. Cathy Caruth, *Unclaimed Experience: Trauma, Narrative, and History* (Baltimore, MD and London: Johns Hopkins University Press, 1996), 8. See also chapter 5, "Traumatic Awakenings (Freud, Lacan, and the Ethics of Memory)," 91–112.

7. Charles Darwin, *On the Origin of Species by Means of Natural Selection, or the Preservation of Favoured Races in the Struggle for Life* (1859; reprint, Cambridge, MA and London: Harvard University Press, 1964), 62.

8. W. G. Sebald, *Austerlitz*, trans. Anthea Bell (New York: Modern Library, 2001), 24. Subsequent citations appear in the text following the abbreviation *A*.

9. James K. Chandler, "About Loss: W. G. Sebald's Romantic Art of Memory," *SAQ* 102.1 (Winter 2003): 258.

10. See also the photograph on this page of a cabinet containing pinned and labeled butterflies.

11. Austerlitz refers to a section of the ninth chapter of the *Journal of Researches*. It, too, ends with loss:

> One evening, when we were about ten miles from the Bay of San Blas, vast numbers of butterflies, in bands or flocks of countless myriads, extended as far as the eye could range. Even by the aid of a glass it was not possible to see a space free from butterflies. The seamen cried out "it was snowing butterflies," and such in fact was the appearance … Before sunset, a strong breeze sprung up from the north, and this must have been the cause of tens of thousands of the butterflies and other insects having perished. (Charles Darwin, *Journal of Researches into the Natural History and Geology of the Various Countries Visited by H.M.S. Beagle, under the Command of Captain FitzRoy, R.N, from 1832 to 1836* (1839), vols. ii and iii, *The Works of Charles Darwin*, 29 vols. (London: Pickering, 1986–89), ii: 145–46)

12. Chandler observes about Sebald's earlier *The Rings of Saturn* that it "concludes with a long discussion of the silk industry in East Anglia since the Middle Ages that evolves into an analysis of the silk industry in Germany, and indeed an analysis of the didactic uses to which the Germans put the practice of *breeding* silk worms both before and during the Nazi period." "About Loss," 240–41; emphasis in the original.

13. Charles Darwin, "A Biographical Sketch of an Infant," *Mind* 2 (July 1877): 289–90. The child was William Erasmus Darwin, born December 27, 1839.

14. Charles Darwin, *Charles Darwin's Notebooks, 1836–1844: Geology, Transmutation of Species, Metaphysical Enquiries*, ed. Paul H. Barrett *et al.* (Ithaca, NY: Cornell University Press, 1987), 551. Darwin also presented a mirror to Willy, a chimpanzee at the London Zoological Gardens. On these encounters, see Janet Browne, *Charles Darwin: Voyaging* (New York: Knopf, 1995), 376–77; Randal Keynes, *Annie's Box: Charles Darwin, His Daughter, and Human Evolution* (London: Fourth Estate, 2001), 45–46, 56; and Adrian Desmond and James Moore, *Darwin: The Life of a Tormented Evolutionist* (New York and London: Norton, 1991), 243–44.

15. See Giorgio Agamben, *The Open: Man and Animal*, trans. Kevin Attell (Stanford, CA: Stanford University Press, 2004); Donna Haraway, *The Companion Species Manifesto: Dogs, People, and Significant Otherness* (Chicago: Prickly Paradigm Press, 2003), *Primate Visions: Gender, Race, and Nature in the World of Modern Science* (New York and London: Routledge, 1989), and *When Species Meet* (Minneapolis and London: University of Minnesota Press, 2007); Cary Wolfe, *Animal Rites: American Culture, the Discourse of Species, and Posthumanist Theory* (Chicago and London: University of Chicago Press, 2003) and *Zoontologies: The Question of the Animal* (Minneapolis and London: University of Minnesota Press, 2003).

16. Darwin, *Charles Darwin's Notebooks*, 559.

17. Gillian Beer, *Open Fields: Science in Cultural Encounter* (Oxford: Clarendon, 1996), 127.

18. Darwin, *Descent of Man*, II: 404; W. H. Hudson, *Idle Days in Patagonia* (1893), reprinted as volume XVI of *Collected Works of W. H. Hudson*, 24 vols. (London: J. M. Dent; New York: Dutton, 1923), 40.

19. Adam Phillips, *Darwin's Worms: On Life Stories and Death Stories* (London: Basic Books, 2000), 14.

Bibliography

Ackerman, Diane, *A Natural History of the Senses* (1990; reprint, New York: Vintage, 1995).

Adams, James Eli, *Dandies and Desert Saints: Styles of Victorian Masculinity* (Ithaca, NY and London: Cornell University Press, 1995).

Agamben, Giorgio. *The Open: Man and Animal*, trans. Kevin Attell (Stanford, CA: Stanford University Press, 2004).

Aguirre, Robert, *Informal Empire: Mexico and Central America in Victorian Culture* (Minneapolis and London: University of Minnesota Press, 2005).

Aguirre, Robert, and Ross Forman, eds., *Connecting Continents: Britain and Latin America, 1780–1900* (New York and Amsterdam: Rodopi, forthcoming).

Alderson, David, *Mansex Fine: Religion, Manliness and Imperialism in Nineteenth-Century British Culture* (Manchester: Manchester University Press, 1998).

Allen, David Elliston, *The Naturalist in Britain: A Social History* (1976; reprint, Harmondsworth: Penguin, 1978).

Allen, Grant, "A Freak of Memory," *The Queen, the Lady's Newspaper* **102** (November 13, 1897): 909–11.

"Introduction," *In the Guiana Forest: Studies of Nature in Relation to the Struggle for Life*, by James Rodway (London: T. Fisher Unwin, 1894), vii–xxiii.

Alston, A. H. G., "Henry Walter Bates: A Centenary," *The Geographical Journal* **112** (July–September 1948): 1–3.

Amigoni, David, *Colonies, Cults, and Evolution: Literature, Science and Culture in Nineteenth-Century Writing* (Cambridge and New York: Cambridge University Press, 2007).

Armitage, David, *The Ideological Origins of the British Empire* (Cambridge and New York: Cambridge University Press, 2000).

Bagehot, Walter, *Physics and Politics; or, Thoughts on the Application of the Principles of "Natural Selection" and "Inheritance" to Political Society* (New York: D. Appleton and Company, 1873).

Barber, Lynn, *The Heyday of Natural History, 1820–1870* (London: Cape, 1980).

Barker, Charles, "Erotic Martyrdom: Kingsley's Sexuality Beyond Sex," *Victorian Studies* **44.3** (Spring 2002): 465–88.

Barrish, Phillip, "Accumulating Variation: Darwin's *On the Origin of Species* and Contemporary Literary and Cultural Theory," *Victorian Studies* **34.3** (Summer 1991): 431–53.

Bates, Henry Walter, Pocket-Book kept while in Amazon (1848–59), BL Add. MS 42138A-B, folios 24–25.

    The Naturalist on the River Amazons, A Record of Adventures, Habits of Animals, Sketches of Brazilian and Indian Life, and Aspects of Nature under the Equator, During Eleven Years of Travel, 2 vols. (London: John Murray, 1863).

Baucom, Ian, *Out of Place: Englishness, Empire, and the Locations of Identity* (Princeton, NJ: Princeton University Press, 1999).

Beaglehole, J.C., *The Life of Captain James Cook* (Stanford, CA: Stanford University Press, 1974).

Beddall, Barbara G., "Wallace, Darwin, and the Theory of Natural Selection: A Study in the Development of Ideas and Attitudes," *Journal of the History of Biology* **5**.1 (1968): 261–323.

Beer, Gillian, *Darwin's Plots: Evolutionary Narrative in Darwin, George Eliot and Nineteenth-Century Fiction* (1983; revised edition, Cambridge and New York: Cambridge University Press, 2000).

    Open Fields: Science in Cultural Encounter (Oxford: Clarendon, 1996).

    "Origins and Oblivion in Victorian Narrative," in *Sex, Politics, and Science in the Nineteenth-Century Novel*, ed. Ruth Bernard Yeazell (Baltimore, MD: Johns Hopkins University Press, 1986), 63–87.

Behdad, Ali, *Belated Travelers: Orientalism in the Age of Colonial Dissolution* (Durham, NC and London: Duke University Press, 1994).

    A Forgetful Nation: On Immigration and Cultural Identity in the United States (Durham, NC and London: Duke University Press, 2005).

Bell, C. Napier, *Tangweera: Life and Adventures among Gentle Savages* (1899; reprint, Austin: University of Texas Press, 1989).

Benjamin, Walter, *Illuminations*, trans. Harry Zohn (New York: Schocken, 1968).

Bergson, Henri, *Matter and Memory* (1896; fifth edition, 1908), trans. N. M. Paul and W. S. Palmer (New York: Zone, 1991).

Bethell, Leslie, *The Abolition of the Brazilian Slave Trade: Britain, Brazil, and the Slave Trade Question, 1807–1869* (Cambridge: Cambridge University Press, 1970).

    "Britain and Latin America in Historical Perspective," in *Britain and Latin America: A Changing Relationship*, ed. Victor Bulmer-Thomas (Cambridge and New York: Cambridge University Press, 1989), 1–24.

Bewell, Alan, *Romanticism and Colonial Disease* (Baltimore, MD: Johns Hopkins University Press, 2000).

Bongie, Chris, *Exotic Memories: Literature, Colonialism, and the Fin de Siècle* (Stanford, CA: Stanford University Press, 1991).

Borges, Jorge Luis, *Other Inquisitions 1937–1952*, trans. Ruth L. C. Simms (1952; English translation, Austin: University of Texas Press, 1964).

Bowen, H. V., *The Business of Empire: The East India Company and Imperial Britain, 1756–1833* (Cambridge and New York: Cambridge University Press, 2006).

Bowler, Peter, *Fossils and Progress: Paleontology and the Idea of Progressive Evolution in the Nineteenth Century* (New York: Science History Publications, 1976).

The Invention of Progress: The Victorians and the Past (Oxford and New York: Blackwell, 1990).

Theories of Human Evolution: A Century of Debate, 1844–1944 (Baltimore, MD: Johns Hopkins University Press, 1986).

Brackman, Arnold C., *A Delicate Arrangement: The Strange Case of Charles Darwin and Alfred Russel Wallace* (New York: Columbia University Press, 1980).

Brantlinger, Patrick, *Dark Vanishings: Discourse on the Extinction of Primitive Races, 1800–1930* (Ithaca, NY and London: Cornell University Press, 2003).

Bridges, Lucas, *Uttermost Part of the Earth* (New York: Dutton, 1949).

Brock, William H. "Humboldt and the British: A Note on the Character of British Science," *Annals of Science 50* (Basingstoke: Taylor and Francis, 1993): 365–72.

Brooks, John Langdon, *Just before the Origin: Alfred Russel Wallace's Theory of Evolution* (New York: Columbia University Press, 1984).

Browne, Janet, *Charles Darwin: The Power of Place* (Princeton, NJ and Oxford: Princeton University Press, 2002).

Charles Darwin: Voyaging (New York: Knopf, 1995).

"Darwin in Caricature: A Study in the Popularisation and Dissemination of Evolution," *Proceedings of the American Philosophical Society* **145.4** (December 2001): 496–509.

Browne, Janet, and Sharon Messenger, "Victorian Spectacle: Julia Pastrana, the Bearded and Hairy Female," *Endeavour* **27.4** (December 2003): 155–59.

Burkhardt, Frederick, Sydney Smith, *et al.*, eds., *The Correspondence of Charles Darwin*, 16 vols. (Cambridge: Cambridge University Press, 1985–).

Burnett, D. Graham, "'It Is Impossible to Make a Step without the Indians': Nineteenth-Century Geographical Exploration and the Amerindians of British Guiana," *Ethnohistory* **49.1** (Winter 2002): 3–40.

Masters of All They Surveyed: Exploration, Geography, and a British El Dorado (Chicago and London: University of Chicago Press, 2000).

Butler, Samuel, *Unconscious Memory* (1880; reprint, New York: Dutton, 1911).

Cain, P. J., and A. G. Hopkins, *British Imperialism: Innovation and Expansion 1688–1914* (London and New York: Longman, 1993).

Camerini, Jane R., "Evolution, Biogeography, and Maps: An Early History of Wallace's Line," *Isis* **84.4** (December 1993): 700–27.

"Remains of the Day: Early Victorians in the Field," in *Victorian Science in Context*, ed. Bernard Lightman (Chicago: University of Chicago Press, 1997), 354–77.

"Wallace in the Field," *Osiris*, second series, **11** (1996): 44–65.

Campbell, Matthew, Jacqueline M. Labbé, and Sally Shuttleworth, eds., *Memory and Memorials 1789–1914: Literary and Cultural Perspectives* (London and New York: Routledge, 2000).

Canon, Susan Faye, *Science in Culture: The Early Victorian Period* (New York: Science History Publications, 1978).

Carruthers, Mary, *The Book of Memory: A Study of Memory in Medieval Culture* (Cambridge: Cambridge University Press, 1990).

Caruth, Cathy, *Unclaimed Experience: Trauma, Narrative and History* (Baltimore, MD and London: Johns Hopkins University Press, 1996).

Caruth, Cathy, ed., *Trauma: Explorations in Memory* (Baltimore, MD: Johns Hopkins University Press, 1995).

Chandler, James K., "About Loss: W. G. Sebald's Romantic Art of Memory," *SAQ* **102.1** (Winter 2003): 235–62.

Wordsworth's Second Nature: A Study of the Poetry and the Politics (Chicago and London: University of Chicago Press, 1984).

Chapman, Anne Mackaye, *Cape Horn: Encounters with the Native People Before and After Darwin* (forthcoming).

Darwin in Tierra del Fuego (Buenos Aires: Imago Mundi, 2006).

Chitty, Susan, *The Beast and the Monk: A Life of Charles Kingsley* (New York: Mason/Charter, 1975).

Colley, Ann C., *Nostalgia and Recollection in Victorian Culture* (Basingstoke and London: Macmillan; New York: St. Martin's, 1998).

Colley, Linda, *Britons: Forging the Nation, 1707–1837* (New Haven, CT and London: Yale University Press, 1992).

Collier, Simon, "The Four Worlds of W.H. Hudson," in *The Borges Tradition*, ed. Norman Thomas di Giovanni (London: Constable, in association with the Anglo-Argentine Society, 1995), 71–88.

Colls, Robert, and Philip Dodd, eds., *Englishness: Politics and Culture, 1880–1920* (London: Croom Helm, 1986).

Connerton, Paul, *How Societies Remember* (Cambridge and New York: Cambridge University Press, 1989).

Conrad, Joseph, *Heart of Darkness* (1899; reprint, Harmondsworth: Penguin, 1995).

Corbin, Alain, *Time, Desire and Horror: Towards a History of the Senses*, trans. Jean Birrell (Cambridge: Polity, 1995).

Dalziel, Rosamund, "The Curious Case of Sir Everard im Thurn and Sir Arthur Conan Doyle: Exploration and the Imperial Adventure Novel, *The Lost World*," *ELT* **45.2** (2002): 131–57.

Dames, Nicholas, *Amnesiac Selves: Nostalgia, Forgetting, and British Fiction, 1810–1870* (Oxford and New York: Oxford University Press, 2001).

Darwin, Charles, *The Autobiography of Charles Darwin, 1809–1882* (1887; reprint, New York and London: Norton, 1969).

"A Biographical Sketch of an Infant," *Mind* **2** (July 1877): 285–94.

Charles Darwin's Notebooks, 1836–1844: Geology, Transmutation of Species, Metaphysical Enquiries, ed. Paul H. Barrett *et al.* (Ithaca, NY: Cornell University Press, 1987).

The Descent of Man, and Selection in Relation to Sex (1871; reprint, Princeton, NJ: Princeton University Press, 1981).

The Formation of Vegetable Mould, Through the Action of Worms, with Observations on Their Habits (1881; reprint, Chicago and London: University of Chicago Press, 1985).

On the Origin of Species by Means of Natural Selection, or the Preservation of Favoured Races in the Struggle for Life (1859; reprint, Cambridge, MA and London: Harvard University Press, 1964).

On the Origin of Species by Means of Natural Selection, or the Preservation of Favoured Races in the Struggle for Life (1859; second edition, London: John Murray, 1860),

The Works of Charles Darwin, 29 vols. (London: Pickering, 1986–89).

Darwin, Francis, ed., *The Life and Letters of Charles Darwin, Including an Autobiographical Chapter*, 3 vols. (London: John Murray, 1887).

Daston, Lorraine, "Type Specimens and Scientific Memory," *Critical Inquiry* **31** (Autumn 2004): 153–82.

Dawson, Gowan, *Darwin, Literature and Victorian Respectability* (Cambridge and New York: Cambridge University Press, 2007).

Deleuze, Gilles, *Bergsonism*, trans. Hugh Tomlinson and Barbara Habberjam (1966; New York: Zone, 1991).

Desmond, Adrian, and James Moore, *Darwin: The Life of a Tormented Evolutionist* (New York and London: Norton, 1991).

Dilke, Charles Wentworth, *Greater Britain: A Record of Travel in English-Speaking Countries During 1866–7*, 2 vols. (1868; reprint, London: Macmillan, 1869).

Döring, Tobias, "The Sea is History: Historicizing the Homeric Sea in Victorian Passages," in *Fictions of the Sea: Critical Perspectives on the Ocean in British Literature and Culture*, ed. Bernhard Klein (Aldershot, Hampshire and Burlington, VT: Ashgate, 2002), 121–40.

Doyle, Arthur Conan, *The Lost World* (1912), reprinted in *The Lost World and Other Stories* (Ware: Wordsworth, 1995), 1–169.

Drayton, Richard, *Nature's Government: Science, Imperial Britain, and the "Improvement" of the World* (New Haven, CT and London: Yale University Press, 2000).

Dumett, Raymond E., ed., *Gentlemanly Capitalism and British Imperialism: The New Debate on Empire* (London and New York: Longman, 1999).

Duncan, Ian, "Darwin and the Savages," *Yale Journal of Criticism* **4** (1991): 13–45.

"Introduction," *Green Mansions: A Romance of the Tropical Forest*, by W. H. Hudson (1904; reprint, Oxford and New York: Oxford University Press, 1998), vii–xxiii.

Durant, John, "The Ascent of Nature in Darwin's *Descent of Man*," in *The Darwinian Heritage*, ed. David Kohn (Princeton, NJ: Princeton University Press, 1985), 283–306.

"Scientific Naturalism and Social Reform in the Thought of Alfred Russel Wallace," *The British Journal for the History of Science* **12** (1979): 31–58.

Elliott, Michael A., "Other Times: Herman Melville, Lewis Henry Morgan, and Ethnographic Writing in the Antebellum United States," *Criticism* **49.4** (Fall 2007): 481–503.

Endersby, Jim, *Imperial Nature: Joseph Hooker and the Practices of Victorian Science* (Chicago and London: University of Chicago Press, 2008).

Engels, Friedrich, *Anti-Dühring* (1877; reprint, Beijing: Foreign Languages Press, 1976).

Evans, Brad, *Before Cultures: The Ethnographic Imagination in American Literature, 1865–1920* (Chicago and London: University of Chicago Press, 2005).

Fabian, Johannes, *Time and the Other: How Anthropology Makes Its Object* (New York: Columbia University Press, 1983).

Fan, Fa-ti, *British Naturalists in Qing China: Science, Empire, and Cultural Encounter* (Cambridge, MA and London: Harvard University Press, 2004).

Ferguson, Frances, "Romantic Memory," *Studies in Romanticism* **35**.4 (1996): 509–33.

Fichman, Martin, *An Elusive Victorian: The Evolution of Alfred Russel Wallace* (Chicago and London: University of Chicago Press, *2004*).

Evolutionary Theory and Victorian Culture (Amherst, NY: Humanity, 2002).

Fitzroy, Robert, *Proceedings of the Second Expedition, 1831–1836, under the Command of Captain Robert Fitz-Roy* [*sic*], vol. II, *Narrative of the Surveying Voyages of H.M.S. Adventure and Beagle, between the Years 1826 and 1836, Describing Their Examination of the Southern Shores of South America, and the Beagle's Circumnavigation of the Globe*, ed. Fitzroy (London: Henry Colburn, 1839).

Flint, Kate, "Sensuous Knowledge," in *Unmapped Countries: Biological Visions in Nineteenth Century Literature and Culture*, ed. Anne-Julia Zwierlein (London: Anthem, 2005), 207–15.

Ford, Ford Madox, *Mightier than the Sword* (London: Allen & Unwin, 1938).

Forman, Ross, "Projecting from Possession Point: Hong Kong, Hybridity, and the Shifting Grounds of Imperialism in James Dalziel's Turn-of-the-Century Fiction," *Criticism* **46.4** (Fall 2004): 533–74.

Foucault, Michel, *The Order of Things: An Archaeology of the Human Sciences* (1966; English translation, New York: Pantheon, 1970).

Freud, Sigmund, "A Difficulty in the Path of Psychoanalysis" (1917), reprinted in *The Standard Edition of the Complete Psychological Works of Sigmund Freud*, trans. James Strachey *et al.*, 24 vols. (London: Hogarth Press, 1955–74), XVII: 136–44.

Froude, James Anthony, *The English in the West Indies; or, The Bow of Ulysses* (London: Longmans, Green, and Co., 1888).

Frow, John, *Time and Commodity Culture: Essays in Cultural Theory and Postmodernity* (Oxford: Clarendon, 1997).

Fulford, Tim, Debbie Lee, and Peter J. Kitson, *Literature, Science and Exploration in the Romantic Era: Bodies of Knowledge* (Cambridge and New York: Cambridge University Press, 2004).

Gallagher, John, and Ronald Robinson, "The Imperialism of Free Trade," *Economic History Review* **6** (1953): 1–15.

Garnett, Edward, "The Genius of W. H. Hudson," in *A Hind in Richmond Park*, by W. H. Hudson (New York: Dutton, 1923), xiii–xxii.

Gates, Barbara T., *Kindred Nature: Victorian and Edwardian Women Embrace the Living World* (Chicago: University of Chicago Press, 1999).

Genette, Gérard, *Narrative Discourse: An Essay in Method*, trans. Jane E. Lewin (1972; English translation, Ithaca, NY: Cornell University Press, 1980).

George, Wilma B., *Biologist Philosopher: A Study of the Life and Writings of Alfred Russel Wallace* (London and New York: Abelard-Schuman, 1964).

Gikandi, Simon, *Maps of Englishness: Writing Identity in the Culture of Colonialism* (New York: Columbia University Press, 1996).

Glade, William, "Latin America and the International Economy, 1870–1914," in *The Cambridge History of Latin America*, ed. Leslie Bethell, 11 vols. (Cambridge: Cambridge University Press, 1984–94), IV: 1–56.

Glendening, John, "Darwinian Entanglement in Hudson's *Green Mansions*," *ELT* **43** (2000): 259–79.

Goddard, Harold, *W. H. Hudson: Bird-Man* (New York: Dutton, 1928).

GoGwilt, Christopher, *The Invention of the West: Joseph Conrad and the Double-Mapping of Europe and Empire* (Stanford, CA: Stanford University Press, 1995).

Gosse, Edmund, *Father and Son* (1907; reprint, Harmondsworth: Penguin, 1983).

Gould, Stephen Jay, *The Mismeasure of Man* (New York: Norton, 1981).

Ontogeny and Phylogeny (Cambridge, MA and London: Harvard University Press, 1977).

Time's Arrow, Time's Cycle: Myth and Metaphor in the Discovery of Geological Time (Cambridge, MA and London: Harvard University Press, 1987).

Wonderful Life: The Burgess Shale and the Nature of History (New York and London: Norton, 1989).

Graham, G. S., and R. A. Humphreys, eds., *The Navy and South America, 1807–23* (Cambridge: Cambridge University Press, 1962).

Grayson, Donald, ed., *The Establishment of Human Antiquity* (New York: Academic, 1983).

Grene, Marjorie, and David Depew, *The Philosophy of Biology: An Episodic History* (Cambridge and New York: Cambridge University Press, 2004).

Gross, John, "Editor's Introduction," *The Expansion of England*, by Sir John R. Seeley (1883; reprint, Chicago and London: University of Chicago Press, 1971), xi–xxviii.

Habermas, Jürgen, *The Philosophical Discourse of Modernity: Twelve Lectures*, trans. Frederick Lawrence (Cambridge, MA: MIT Press, 1987).

Hacking, Ian, *Rewriting the Soul: Multiple Personality and the Sciences of Memory* (Princeton, NJ: Princeton University Press, 1995).

Haeckel, Ernst, *The Evolution of Man: A Popular Exposition of the Principal Points of Human Ontogeny and Phylogeny*, 2 vols. (1874; English translation of third edition, New York: D. Appleton and Co., 1879).

Halbwachs, Maurice, *On Collective Memory*, trans. Lewis Coser (Chicago: University of Chicago Press, 1992).

Hall, Catherine, *Civilising Subjects: Metropole and Colony in the English Imagination 1830–1867* (Chicago and London: University of Chicago Press, 2002).

White, Male, and Middle-Class: Explorations in Feminism and History (Cambridge: Polity, 1992).

Hall, Catherine, ed., *Cultures of Empire: Colonizers in Britain and the Empire in the Nineteenth and Twentieth Centuries. A Reader* (New York: Routledge, 2000).

Hall, Donald E., *Muscular Christianity: Embodying the Victorian Age* (Cambridge: Cambridge University Press, 1994).

Hansen, Miriam, "Benjamin and Cinema: Not a One-Way Street," in *Benjamin's Ghosts: Interventions in Contemporary Literary and Cultural Theory*, ed. Gerhard Richter (Stanford, CA: Stanford University Press, 2002), 41–73.

Haraway, Donna, *The Companion Species Manifesto: Dogs, People, and Significant Otherness* (Chicago: Prickly Paradigm Press, 2003).

Primate Visions: Gender, Race, and Nature in the World of Modern Science (New York and London: Routledge, 1989).

When Species Meet (Minneapolis and London: University of Minnesota Press, 2007).

Hazlewood, Nick, *Savage: The Life and Times of Jemmy Button* (London: Hodder and Stoughton, 2000).

Henry, Nancy, and Cannon Schmitt, eds., *Victorian Investments: New Perspectives on Culture and Finance* (Bloomington and Indianapolis: Indiana University Press, 2008).

Herbert, Christopher, *Culture and Anomie: Ethnographic Imagination in the Nineteenth Century* (Chicago and London: University of Chicago Press, 1991).

Herbert, Sandra, *Charles Darwin, Geologist* (Ithaca, NY and London: Cornell University Press, 2005).

"The Place of Man in the Development of Darwin's Theory of Transmutation, Part I. To July 1837," *Journal of the History of Biology* **7.2** (Fall 1974): 217–58.

Heringman, Noah, *Romantic Rocks, Aesthetic Geology* (Ithaca, NY and London: Cornell University Press, 2004).

Hoad, Neville, "Wild(e) Men and Savages: The Homosexual and the Primitive in Darwin, Wilde and Freud" (Ph.D. diss., Columbia University, 1998).

Hudson, W. H., *Collected Works of W. H. Hudson*, 24 vols. (London: J. M. Dent; New York: Dutton, 1923).

Far Away and Long Ago: A Childhood in Argentina (1918; reprint, London: Eland Books, 1982).

Green Mansions: A Romance of the Tropical Forest (1904; reprint, Oxford and New York: Oxford University Press, 1998).

A Hind in Richmond Park (New York: Dutton, 1923).

Letter to the Secretary of the Zoological Society, London, *Proceedings of the Zoological Society*, Part I (1870): 158–60.

The Naturalist in La Plata (London: Chapman and Hall, 1892).

The Unpublished Letters of W. H. Hudson, The First Literary Environmentalist 1841–1922, ed. Dennis Shrubsall, 2 vols. (Lewiston, NY, Queenston, ON, and Lampeter, Ceredigion: The Edwin Mellen Press, 2006).

Hulme, Peter, "Cast Away: The Uttermost Parts of the Earth," in *Sea Changes: Historicizing the Ocean*, ed. Bernhard Klein and Gesa Mackenthun (New York and London: Routledge, 2004), 187–201.

Huxley, Thomas Henry, *Collected Essays*, 9 vols. (New York: Appleton, 1894–98).

Critiques and Essays (New York: Appleton, 1887).

Evidence as to Man's Place in Nature (London and Edinburgh: Williams and Norgate, 1863).

"Science at Sea," *Westminster Review* **61** (1854): 98–119.

T. H. Huxley's Diary of the Voyage of H. M. S. Rattlesnake, ed. Julian Huxley (Garden City, NY: Doubleday, 1931).

Huyssen, Andreas, *Present Pasts: Urban Palimpsests and the Politics of Memory* (Stanford, CA: Stanford University Press, 2003).

Twilight Memories: Marking Time in a Culture of Amnesia (London: Routledge, 1995).

im Thurn, Everard F., *Among the Indians of Guiana, being Sketches Chiefly Anthropologic from the Interior of British Guiana* (London: Kegan, Paul, Trench and Co., 1883).

Jagoe, Eva-Lynn Alicia, *The End of the World as They Knew It: Writing Experiences of the Argentine South* (Lewisburg, PA: Bucknell University Press, 2008).

James, William, *The Principles of Psychology* (1890; reprint, Cambridge, MA: Harvard University Press, 1981).

Jardine, Nicholas, J. A. Secord, and E. C. Spary, eds., *Cultures of Natural History* (Cambridge: Cambridge University Press, 1996).

Jones, Ann Rosalind, and Peter Stallybrass, *Renaissance Clothing and the Materials of Memory* (Cambridge and New York: Cambridge University Press, 2000).

Kaufmann, William W., *British Policy and the Independence of Latin America, 1804–1828* (New Haven, CT: Yale University Press, 1951).

Kendall, Guy, *Charles Kingsley and His Ideas* (London: Hutchinson and Co., n.d.).

Keynes, Randal, *Annie's Box: Charles Darwin, His Daughter, and Human Evolution* (London: Fourth Estate, 2001).

Keynes, Richard Darwin, *Fossils, Finches and Fuegians: Darwin's Adventures and Discoveries on the Beagle* (Oxford and New York: Oxford University Press, 2003).

King, P. Parker, *Proceedings of the First Expedition, 1826–1830, under the Command of Captain P. Parker King*, vol. 1, *Narrative of the Surveying Voyages of H.M.S. Adventure and Beagle, between the Years 1826 and 1836, Describing Their Examination of the Southern Shores of South America, and the Beagle's Circumnavigation of the Globe*, ed. Robert Fitzroy (London: Henry Colburn, 1839).

Kingsley, Charles, *Glaucus; or, The Wonders of the Shore* (Boston: Ticknor and Fields, 1855).

Scientific Lectures and Essays (1880; revised edition, London and New York: Macmillan and Co., 1890).

The Works of Charles Kingsley, 28 vols. (London: Macmillan, 1880–85).

Klaver, J. M. I., *The Apostle of the Flesh: A Critical Life of Charles Kingsley* (Leiden and Boston: Brill, 2006).

Knight, Alan, "Britain and Latin America," in *The Oxford History of the British Empire: The Nineteenth Century*, ed. Andrew Porter, 5 vols. (Oxford: Oxford University Press, 1999), III: 122–45.

Koselleck, Reinhart, *Futures Past: On the Semantics of Historical Time*, trans. Keith Tribe (Cambridge, MA and London: MIT Press, 1985).

Kreilkamp, Ivan, *Voice and the Victorian Storyteller* (Cambridge and New York: Cambridge University Press, 2005).

Kuhn, Thomas, *The Structure of Scientific Revolutions* (1962; third edition, Chicago and London: University of Chicago Press, 1996).

Kuper, Adam, "Mirror, Mirror," *Criticism* **49.4** (Fall 2007): 551–56.

Lamb, Jonathan, "Metamorphosis and Settlement: The Enlightened Anthropology of Colonial Societies," in *The Anthropology of the Enlightenment*, ed. Larry Wolff and Marco Cipolloni (Stanford, CA: Stanford University Press, 2007), 277–91.

 Preserving the Self in the South Seas, 1680–1840 (Chicago and London: University of Chicago Press, 2001).

Landsberg, Alison, *Prosthetic Memory: The Transformation of American Remembrance in the Age of Mass Culture* (New York: Columbia University Press, 2004).

Latour, Bruno, *We Have Never Been Modern* (Cambridge, MA: Harvard University Press, 1993).

Le Goff, Jacques, *History and Memory*, trans. Steven Rendall and Elizabeth Claman (New York: Columbia University Press, 1992).

Leask, Nigel, "Darwin's 'Second Sun': Alexander von Humboldt and the Genesis of *The Voyage of the Beagle*," in *Literature, Science, Psychoanalysis, 1830–1970: Essays in Honor of Gillian Beer*, ed. Helen Small and Trudi Tate (Oxford and New York: Oxford University Press, 2003), 13–36.

Leith, James A., review of Pierre Nora, ed., *Realms of Memory: the Construction of the French Past*, vol. III, *The Symbols*, English language edition ed. Laurence D. Kritzman, trans. Arthur Goldhammer (New York and Chichester, West Sussex: Columbia University Press, 1998); www.h-france.net/reviews/leith.html.

Levine, George, *Darwin Loves You: Natural Selection and the Re-enchantment of the World* (Princeton, NJ: Princeton University Press, 2006).

Levine, Philippa, "States of Undress: Nakedness and the Colonial Imagination," *Victorian Studies* **50.2** (Winter 2008): 189–219.

Lewes, George Henry, *Studies in Animal Life* (London: Smith, Elder, 1862).

Li, Victor, "A Necessary Vigilance: A Response to Torgovnick and Kuper," *Criticism* **49.4** (Fall 2007): 557–63.

 The Neo-Primitivist Turn: Critical Reflections on Alterity, Culture, and Modernity (Toronto: University of Toronto Press, 2006).

Livingstone, David N., *Putting Science in Its Place: Geographies of Scientific Knowledge* (Chicago and London: University of Chicago Press, 2003).

Loewenberg, Bert James, *Darwin, Wallace and the Theory of Natural Selection; including the Linnean Society Papers* (New Haven, CT: G. E. Cinamon, 1957).

Lorimer, Douglas, *Colour, Class, and the Victorians: English Attitudes to the Negro in the Mid-Nineteenth Century* (Leicester: Leicester University Press, 1978).

Lovejoy, Arthur O., "Prolegomena to the History of Primitivism," in *Primitivism and Related Ideas in Antiquity*, by Lovejoy and George Boas (Baltimore, MD: Johns Hopkins University Press, 1935), 1–22.

Lyell, Charles, *Principles of Geology, Being an Attempt to Explain the Former Changes of the Earth's Surface, by Reference to Causes Now in Operation*, 3 vols. (1830–33; reprint, Chicago: University of Chicago Press, 1991).

Magee, Paul, *From Here to Tierra del Fuego* (Urbana and Chicago: University of Illinois Press, 2000).

Marsden, Barry M., *Pioneers of Prehistory: Leaders and Landmarks in English Archaeology, 1500–1900* (Ormskirk: Hesketh, 1984).

Marshall, Nancy Rose, "'A Dim World, Where Monsters Dwell': The Spatial Time of the Sydenham Crystal Palace Dinosaur Park," *Victorian Studies* **49.2** (Winter 2007): 286–301.

Martin, Robert Bernard, *The Dust of Combat: A Life of Charles Kingsley* (New York: Norton, 1960).

Mason, Michael, "The Cultivation of the Senses for Creative Nostalgia in the Essays of W. H. Hudson," *ARIEL: A Review of International English Literature* **20** (1989): 23–37.

Mason, Peter, *Infelicities: Representations of the Exotic* (Baltimore, MD and London: Johns Hopkins University Press, 1998).

The Lives of Images (London: Reaktion, 2001).

Matus, Jill, "Trauma, Memory, and the Railway Disaster: The Dickensian Connection," *Victorian Studies* **43.3** (Spring 2001): 413–36.

Maynard, John, *Victorian Discourses on Sexuality and Religion* (Cambridge: Cambridge University Press, 1993).

McClintock, Anne, *Imperial Leather: Race, Gender, and Sexuality in the Colonial Contest* (New York and London: Routledge, 1995).

McKinney, H. Lewis, *Wallace and Natural Selection* (New Haven, CT and London: Yale University Press, 1972).

Mercer, Kobena, *Welcome to the Jungle: New Positions in Black Cultural Studies* (New York and London: Routledge, 1994).

Merrill, Lynn L., *The Romance of Victorian Natural History* (New York: Oxford University Press, 1989).

Michaels, Walter Benn, "'You who never was there': Slavery and the New Historicism, Deconstruction and the Holocaust," *Narrative* **4** (1996): 1–16.

Mill, J. S., *Dissertations and Discussions, Political, Philosophical, and Historical* (third edition; London: Longmans, Green, Reader, and Dyer, 1875).

Miller, David, *W. H. Hudson and the Elusive Paradise* (New York: St. Martin's, 1990).

Moore, James, "Wallace's Malthusian Moment: The Common Context Revisited," in *Victorian Science in Context*, ed. Bernard Lightman (Chicago: University of Chicago Press, 1997), 290–311.

Moore, James, and Adrian Desmond, "Introduction," in *The Descent of Man, and Selection in Relation to Sex*, by Charles Darwin (London: Penguin, 2004), xi–lviii.

Moorehead, Alan, *Darwin and the Beagle* (London: Hamilton, 1969).

Naipaul, V. S., *The Loss of El Dorado: A History* (New York: Knopf, 1970).

Nairn, Tom, *The Break-up of Britain: Crisis and Neo-nationalism* (London: New Left Books, 1977).

Nash, Roderick, "The Export and Import of Nature," *Perspectives in American History* **12** (1979): 519–60.

Naylor, Bernard, *Accounts of Nineteenth-Century South America* (London: Athlone, 1969).

Nora, Pierre, "Between Memory and History: *Les Lieux de Mémoire*." *Representations* **26** (1989): 7–24.

Osterhammel, Jürgen, "Semi-Colonialism and Informal Empire in Twentieth-Century China: Towards a Framework of Analysis," in *Imperialism and After: Continuities and Discontinuities*, ed. Wolfgang J. Mommsen and Osterhammel (London: German Historical Institute/Allen & Unwin, 1986), 290–314.

Otis, Laura, *Organic Memory: History and the Body in the Late Nineteenth and Early Twentieth Centuries* (Lincoln: University of Nebraska Press, 1994).

Pagden, Anthony, "The Empire's New Clothes: From Empire to Federation, Yesterday and Today," *Common Knowledge* **12.1** (Winter 2006): 44–45.

"The Palms of Tropical America," in *Illustrated Travels: A Record of Discovery, Geography, and Adventure*, ed. Henry Walter Bates, 2 vols. (London: Cassell, Petter, and Galpin, n.d.), II: 127–28.

Pascoe, Judith, *The Hummingbird Cabinet: A Rare and Curious History of Romantic Collectors* (Ithaca, NY and London: Cornell University Press, 2005).

Philip, Kavita, "Imperial Science Rescues a Tree: Global Botanic Networks, Local Knowledge and the Transcontinental Transplantation of Cinchona," *Environment and History* **1** (1995): 173–200.

Phillips, Adam, *Darwin's Worms: On Life Stories and Death Stories* (New York: Basic Books, 2000).

Phillips, Forbes, "Ancestral Memory: A Suggestion," *Nineteenth Century* **59** (1906): 977–83.

Platt, D. C. M., ed., *Business Imperialism, 1840–1930: An Inquiry Based on British Experience in Latin America* (Oxford: Clarendon, 1977).

Pound, Ezra, "Hudson: Poet Strayed into Science," in *Selected Prose, 1909–1965* (London: Faber and Faber, 1978), 399–402.

Pratt, Mary Louise, *Imperial Eyes: Travel Writing and Transculturation* (London and New York: Routledge, 1992).

Proust, Marcel, *A la recherche du temps perdu*, 3 vols. (Paris: Editions Gallimard, 1954).

  Swann's Way, trans. C. K. Scott Moncrieff (1913; English translation, Harmondsworth: Penguin, 1957).

Raby, Peter, *Alfred Russel Wallace: A Life* (Princeton, NJ: Princeton University Press, 2001).

Bright Paradise: Victorian Scientific Travellers (London: Chatto and Windus, 1996).

Richards, Robert J., *Darwin and the Emergence of Evolutionary Theories of Mind and Behavior* (Chicago and London: University of Chicago Press, 1987).

"Darwin on Mind, Morals and Emotions," in *The Cambridge Companion to Darwin*, ed. Jonathan Hodge and Gregory Radick (Cambridge and New York: Cambridge University Press, 2003), 92–115.

"The Epistemology of Historical Interpretation: Progressivity and Recapitulation in Darwin's Theory," in *Biology and Epistemology*, ed. Richard Creath and Jane Maienschein (Cambridge and New York: Cambridge University Press, 2000), 64–88.

The Meaning of Evolution: The Morphological Construction and Ideological Reconstruction of Darwin's Theory (Chicago and London: University of Chicago Press, 1992).

The Tragic Sense of Life: Ernst Haeckel and the Struggle over Evolutionary Thought (Chicago and London: University of Chicago Press, 2008).

Ricoeur, Paul, *Memory, History, Forgetting*, trans. Kathleen Blamey and David Pellauer (Chicago and London: University of Chicago Press, 2004).

Ritvo, Harriet, "Zoological Nomenclature and the Empire of Victorian Science," in *Victorian Science in Context*, ed. Bernard Lightman (Chicago: University of Chicago Press, 1997), 334–53.

Rivière, Peter, ed., *The Guiana Travels of Robert Schomburgk 1835–1844, Volume 1: Explorations on Behalf of the Royal Geographical Society 1835–1839* (Aldershot, Hampshire, and Burlington, VT: Ashgate, for the Hakluyt Society, 2006).

Roberts, Morley, *W. H. Hudson: A Portrait* (New York: E. P. Dutton and Company, 1924).

Rodway, James, *In the Guiana Forest: Studies of Nature in Relation to the Struggle for Life* (London: T. Fisher Unwin, 1894).

Rudwick, Martin J.S., *Bursting the Limits of Time: The Reconstruction of Geohistory in the Age of Revolution* (Chicago and London: University of Chicago Press, 2005).

Scenes from Deep Time: Early Pictorial Representations of the Prehistoric World (Chicago and London: University of Chicago Press, 1995).

Rupke, Nicholas, "A Geography of Enlightenment: The Critical Reception of Humboldt's Work," in *Geography and Enlightenment*, ed. David N. Livingstone and Charles Withers (Chicago: University of Chicago Press, 1999), 319–39.

Said, Edward, *Culture and Imperialism* (New York: Knopf, 1993).

Orientalism (1978; reprint, New York: Vintage, 1979).

Sarmiento, Domingo Faustino, *Life in the Argentine Republic in the Days of the Tyrants; or, Civilization and Barbarism*, trans. Mary Mann (first edition, 1845; English translation, from the third Spanish edition, New York: Hurd and Houghton; Cambridge: Riverside Press, 1868).

Schmitt, Cannon, "Rumor, Shares, and Novelistic Form: Joseph Conrad's *Nostromo*," in *Victorian Investments: New Perspectives on Finance and Culture*, ed. Nancy Henry and Schmitt (Bloomington and Indianapolis: Indiana University Press, 2008), 182–201.

——— "'The Sun and Moon Were Made to Give Them Light': Empire in the Victorian Novel," in *The Concise Companion to the Victorian Novel*, ed. Francis O'Gorman (London: Basil Blackwell, 2004), 4–24.

——— "Victorian Beetlemania," in *Victorian Animal Dreams: Representations of Animals in Victorian Literature and Culture*, ed. Deborah Morse and Martin Danahay (Aldershot, Hampshire and Burlington, VT: Ashgate, 2007), 35–51.

Schmitt, Cannon, Nancy Henry, and Anjali Arondekar, eds., "Victorian Investments," special issue, *Victorian Studies* **45.1** (2002).

Schultes, R. E., "Richard Spruce and the Potential for European Settlement of the Amazon: An Unpublished Letter," *Botanical Journal of the Linnean Society* **77** (September 1978): 131–39.

Sebald, W. G., *Austerlitz*, trans. Anthea Bell (New York: Modern Library, 2001).

——— *On the Natural History of Destruction*, trans. Anthea Bell (1999; English translation, New York: Modern Library, 2004).

Secord, James A., *Victorian Sensation: The Extraordinary Publication, Reception, and Secret Authorship of Vestiges of the Natural History of Creation* (Chicago and London: University of Chicago Press, 2000).

Seeley, John R., *The Expansion of England* (1883; reprint, Chicago and London: University of Chicago Press, 1971).

Seitler, Dana, *Atavistic Tendencies: The Culture of Science in American Modernity* (Minneapolis and London: University of Minnesota Press, 2008).

——— "Freud's Menagerie," *Genre* **38** (Spring/Summer 2005): 45–70.

Shakespeare, Nicholas, "Preface" to W. H. Hudson, *Far Away and Long Ago: A Childhood in Argentina* (1918; reprint, London: Eland Books, 1982), xiii–xvii.

Shakespeare, William, *The Comedy of Errors*, in *The Riverside Shakespeare*, ed. G. Blakemore Evans (second edition; Boston: Houghton Mifflin, 1997).

Shermer, Michael, *In Darwin's Shadow: The Life and Science of Alfred Russel Wallace* (Oxford and New York: Touchstone, 2002).

Shoumatoff, Alex, "Introduction," *The Naturalist on the River Amazons*, by Henry Walter Bates (1863; reprint, Harmondsworth: Penguin, 1989), vii–xviii.

Shuttleworth, Sally, "'The Malady of Thought': Embodied Memory in Victorian Psychology and the Novel," in *Memory and Memorials, 1789–1914: Literary and Cultural Perspectives*, ed. Matthew Campbell, Jacqueline M. Labbé, and Shuttleworth (London and New York: Routledge, 2000), 46–59.

Slotten, Ross A., *The Heretic in Darwin's Court: The Life of Alfred Russel Wallace* (New York: Columbia University Press, 2004).

Smith, Anthony, *Explorers of the Amazon* (New York: Viking, 1990).

Smith, Barbara Herrnstein, "Animal Relatives, Difficult Relations," *differences* **15.1** (2004): 1–19.

Smith, Jonathan, *Charles Darwin and Victorian Visual Culture* (Cambridge and New York: Cambridge University Press, 2006).

Smith, Robert Freeman, "Latin America, the United States and the European Powers, 1830–1930," in *The Cambridge History of Latin America*, ed. Leslie Bethell, 11 vols. (Cambridge: Cambridge University Press, 1984–94), IV: 83–119.

Spary, Emma, "Political, Natural and Bodily Economies," in *Cultures of Natural History*, ed. Nicholas Jardine, James A. Secord, and Spary (Cambridge: Cambridge University Press, 1996), 178–96.

Spencer, Herbert, *The Principles of Biology*, 2 vols. (London: Williams and Norgate, 1864).

Spicer, Jakki, "The Author is Dead, Long Live the Author: Autobiography and the Fantasy of the Individual," *Criticism* **47.3** (Summer 2005): 387–403.

Spivak, Gayatri Chakravorty, *A Critique of Postcolonial Reason: Toward a History of the Vanishing Present* (Cambridge, MA and London: Harvard University Press, 1999).

Stack, D.A., "The First Darwinian Left: Radical and Socialist Responses to Darwin, 1859–1914," *History of Political Thought* **21.4** (Winter 2000): 682–710.

Stepan, Nancy Leys, *The Idea of Race in Science: Great Britain 1800–1960* (Hamden, CT: Archon, 1982).

Picturing Tropical Nature (Ithaca, NY: Cornell University Press, 2001).

Stocking, George, *Race, Culture, and Evolution: Essays in the History of Anthropology* (Chicago: University of Chicago Press, 1968).

Victorian Anthropology (New York and London: Free Press, 1987).

Street, Brian, *The Savage in Literature: Representations of "Primitive" Society in English Fiction 1858–1920* (London and Boston: Routledge and Kegan Paul, 1975).

Sulloway, Frank, *Freud: Biologist of the Mind* (New York: Basic Books, 1983).

Taussig, Michael, *Mimesis and Alterity: A Particular History of the Senses* (New York and London: Routledge, 1993).

Terdiman, Richard, *Present Past: Modernity and the Memory Crisis* (Ithaca, NY: Cornell University Press, 1993).

Thomas, J.J., *Froudacity: West Indian Fables by James Anthony Froude* (1888; second edition, London: T. Fisher Unwin, 1889).

Thompson, Keith S., *The Story of Darwin's Ship* (New York: Norton, 1995).

Thomson, Rosemarie Garland, "Narratives of Deviance and Delight: Staring at Julia Pastrana, the 'Extraordinary Lady,'" in *Beyond the Binary: Reconstructing Cultural Identity in a Multicultural Context*, ed. Timothy B. Powell (New Brunswick, NJ: Rutgers University Press, 1999), 81–104.

Thorp, Margaret Farrand, *Charles Kingsley, 1819–1875* (Princeton, NJ: Princeton University Press; London: Oxford University Press, 1937).

Tobin, Beth Fowkes, *Colonizing Nature: The Tropics in British Arts and Letters, 1760–1820* (Philadelphia: University of Pennsylvania Press, 2005).

Tomalin, Ruth, *W. H. Hudson: A Biography* (London: Faber and Faber, 1982).

Torgovnick, Marianna, *Gone Primitive: Savage Intellects, Modern Lives* (Chicago and London: University of Chicago Press, 1990).

"On Victor Li's *The Neo-Primitivist Turn: Critical Reflections on Alterity, Culture, and Modernity*," *Criticism* **49.4** (Fall 2007): 545–50.

Primitive Passions: Men, Women, and the Quest for Ecstasy (Chicago and London: University of Chicago Press, 1996).

Towheed, Shafquat, "The Creative Evolution of Scientific Paradigms: Vernon Lee and the Debate over the Hereditary Transmission of Acquired Characteristics," *Victorian Studies* **49.1** (Autumn 2006): 33–61.

Trollope, Anthony, *The West Indies and the Spanish Main* (second edition, 1860; reprint, London: Cass, 1968).

Turner, Bryan S., "A Note on Nostalgia," *Theory, Culture and Society* **4** (1987): 147–56.

Tylor, E. B., *Primitive Culture: Researches into the Development of Mythology, Philosophy, Religion, Language, Art, and Custom* (London: John Murray, 1871).

Van Riper, A. Bowdoin, *Men Among the Mammoths: Victorian Science and the Discovery of Human Prehistory* (Chicago: University of Chicago Press, 1993).

Vance, Norman, *The Sinews of the Spirit: The Ideal of Christian Manliness in Victorian Literature and Religious Thought* (Cambridge and New York: Cambridge University Press, 1985).

Veblen, Thorstein, *The Theory of the Leisure Class: An Economic Study in the Evolution of Institutions* (1899; reprint, New York: Modern Library, 1934).

Vierna, Angel Guirao de, "Análisis cuantitativo de las expediciones españolas con destino al Nuevo Mundo," in *Ciencia, vida y espacio en Iberoamérica*, ed. José Luis Pestet 3 vols. (Madrid: Consejo Superior de Investigaciones Científicas, 1989), III: 65–93.

Von Hagen, Victor Wolfgang, *South America Called Them: Explorations of the Great Naturalists La Condamine, Humboldt, Darwin, Spruce* (New York: Knopf, 1945).

Wallace, Alfred Russel, *Contributions to the Theory of Natural Selection: A Series of Essays* (London: Macmillan and Co., 1870).

The Malay Archipelago: The Land of the Orang-Utan, and the Bird of Paradise; A Narrative of Travel, with Studies of Man and Nature, 2 vols. (London: Macmillan and Co., 1869).

"A Man of the Time: Dr. Alfred Russel Wallace and His Coming Autobiography," an interview by "J. M.," *The Book Monthly* (May 1905): 548–49.

My Life: A Record of Events and Opinions, 2 vols. (London: Chapman and Hall, 1905).

A Narrative of Travels on the Amazon and Rio Negro, with an Account of the Native Tribes, and Observations on the Climate, Geology, and Natural History of the Amazon Valley (London: Reeve and Co., 1853).

"The Native Problem in South Africa and Elsewhere," *Independent Review* (November 1906): 174–82.

"On the Tendency of Varieties to depart indefinitely from the Original Type," *Journal of the Proceedings of the Linnean Society* **3** (1858): 53–62.

"The Origin of Human Races and the Antiquity of Man Deduced from the Theory of Natural Selection," *Journal of the Anthropological Society of London* **2** (1864): clvii–clxxxvii.

"Preface," *Notes of a Botanist on the Amazon and Andes*, by Richard Spruce, 2 vols. (London: Macmillan, 1908), 1: v–ix.

"A Remarkable Book on the Habits of Animals," *Nature* **45** (April 14, 1892): 553–56.

"Reveries of a Naturalist," *Nature* (March 23, 1893): 483–84.

"Sir Charles Lyell on Geological Climates and the Origin of Species," *Quarterly Review* **126** (April 1869): 359–94.

"Tylor's 'Primitive Culture,' " *The Academy* (February 15, 1872): 69–71.

Wallace, Elizabeth Kowaleski, *The British Slave Trade and Public Memory* (New York: Columbia University Press, 2006).

Webster, C. K., ed., *Britain and the Independence of Latin America, 1812–30: Select Documents*, 2 vols. (London and New York: Oxford University Press, 1938).

Wellek, René, and Austin Warren, *Theory of Literature* (New York: Harcourt, Brace, 1949).

White, Gilbert, *The Natural History and Antiquities of Selborne* (London: Printed by T. Bensley, for B. White and Son, 1789).

Whitehead, Neil L., "South America/Amazonia: The Forest of Marvels," in *The Cambridge Companion to Travel Writing*, ed. Peter Hulme and Tim Youngs (Cambridge: Cambridge University Press, 2003), 123–38.

Wolfe, Cary, *Animal Rites: American Culture, the Discourse of Species, and Posthumanist Theory* (Chicago and London: University of Chicago Press, 2003).

Wolfe, Cary, ed., *Zoontologies: The Question of the Animal* (Minneapolis and London: University of Minnesota Press, 2003).

Wood, Paul B., "The Science of Man," in *Cultures of Natural History*, ed. Nicholas Jardine, James A. Secord, and Emma Spary (Cambridge: Cambridge University Press, 1996), 197–210.

Wordsworth, William, *The Prelude* (1850; reprint, New York and London: W. W. Norton and Company, 1979).

Wormell, Deborah, *Sir John Seeley and the Uses of History* (Cambridge: Cambridge University Press, 1988).

Yates, Frances, *The Art of Memory* (Chicago: University of Chicago Press, 1966).

Young, Robert J.C., *Colonial Desire: Hybridity in Theory, Culture, and Race* (London and New York: Routledge, 1995).

Young, Robert M., *Darwin's Metaphor: Nature's Place in Victorian Culture* (Cambridge: Cambridge University Press, 1985).

Zimmerman, Virginia, *Excavating Victorians* (Albany: State University of New York Press, 2008).

Index

CAMBRIDGE STUDIES IN NINETEENTH-CENTURY
LITERATURE AND CULTURE

General editor
Gillian Beer, *University of Cambridge*

Titles published